Telephoning in English

Third edition

B. Jean Naterop
Rod Revell

CAMBRIDGE
UNIVERSITY PRESS

CAMBRIDGE
UNIVERSITY PRESS

University Printing House, Cambridge CB2 8BS, United Kingdom

One Liberty Plaza, 20th Floor, New York, NY 10006, USA

477 Williamstown Road, Port Melbourne, VIC 3207, Australia

314-321, 3rd Floor, Plot 3, Splendor Forum, Jasola District Centre, New Delhi - 110025, India

79 Anson Road, #06-04/06, Singapore 079906

Cambridge University Press is part of the University of Cambridge.

It furthers the University's mission by disseminating knowledge in the pursuit of education, learning and research at the highest international levels of excellence.

www.cambridge.org
Information on this title: www.cambridge.org/9780521539111

First published 1987
Second edition 1997
Third edition 2004
Reprinted 2019

A catalogue record for this publication is available from the British Library

ISBN 978-0-521-53911-1 Student's Book
ISBN 978-0-521-53912-8 Audio Cassettes (2)
ISBN 978-0-521-53913-5 Audio CDs (2)

Contents

Self-study guide

For learners using the course without a teacher

Aims of the course

The main aim of *Telephoning in English* is to give you practice in understanding and making phone calls in English. The course is for people who are working, or will be working, in business and whose mother tongue is not English. Most of the material gives practice in listening and speaking, but there are also writing exercises, generally in the form of note-taking or message-taking.

What will I learn?

You will learn to *understand* British and American people and people from other parts of the world when they are speaking about normal business matters.
You will learn to *speak* in a way that will help you when you need to make and answer telephone calls.

How do I use *Telephoning in English*?

There is a book, divided into eight units, and a set of two audio cassettes or CDs (Compact Discs).
Each unit consists of three sections: Listening, Language study and Speaking.
The Overview on page VIII gives a breakdown of the contents of each unit.
The following flow chart shows the stages it is necessary to take through each unit.

Listening

Telephone calls

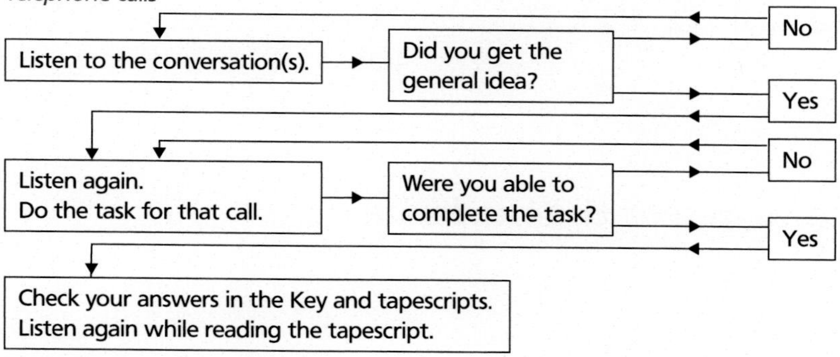

Other activities

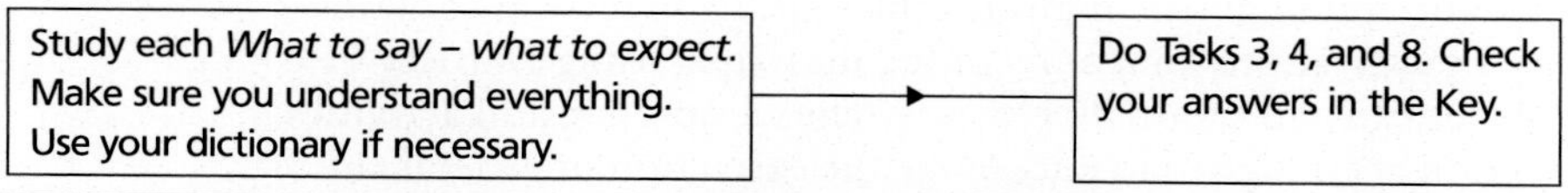

Language study

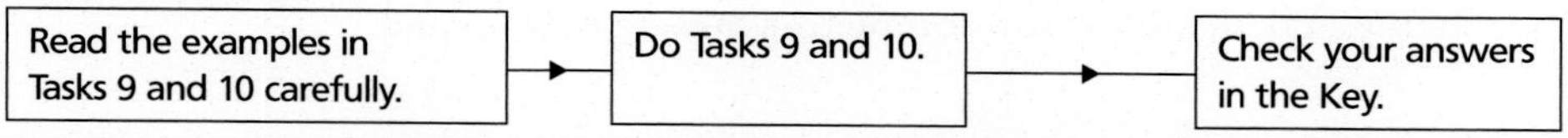

Speaking

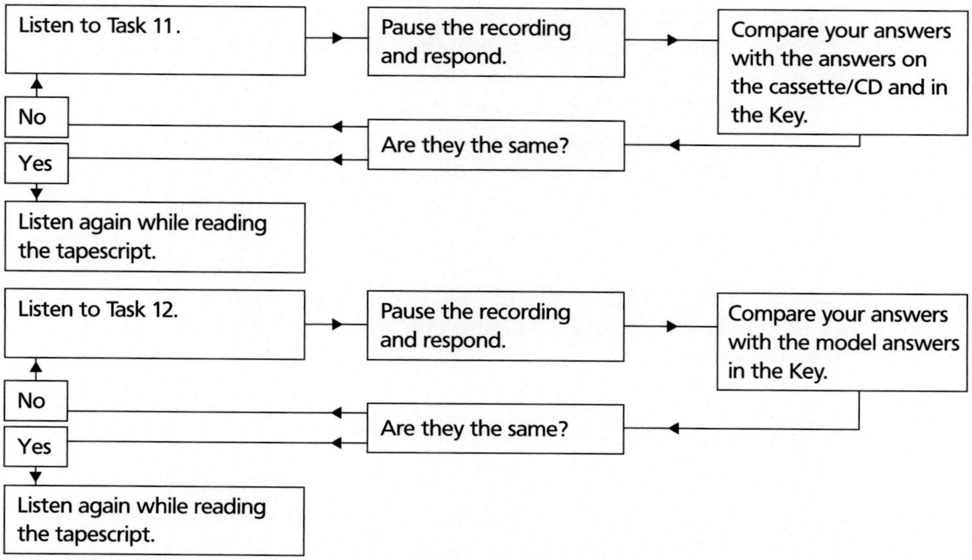

Teacher's notes

Structure and use of the material

Listening

This is the longest section in each of the eight units. It provides the main thematic and linguistic input for the unit. Each section contains telephone conversations and/or recorded messages, using a variety of British, American and non-native voices. The comprehension activities in this section are designed to encourage the extraction of general and detailed information, and to give practice in information recording techniques appropriate to telephone usage. These activities include filling in tables, taking notes, and completing messages, faxes and emails.

What to do

Introduce the conversations briefly. Play them through once without stopping so that the learners can do the comprehension tasks while they listen. If there are two comprehension tasks, play the conversations once more. Discuss the learners' answers with them. At this stage, you may like to play the conversations again and allow the learners to read at the same time in order to confirm their understanding. They should, in any case, not have looked at the text of the conversations before this stage. All the tapescripts are printed at the back of the book (pages 77–114). Between work on the conversations, you should make sure that the learners have studied and know all the 'What to say – what to expect' items. Doing **Tasks 3**, **4** and **8** will also confirm that they can apply what they have learnt.

Language study

A pair of language items that are felt to be appropriate to the type of call being studied and to the learners' level of ability in English are taken out of the listenings for detailed study and exercise. The approach to these items and the way they are exercised are varied.

Introduce each of the language points covered by the exercises in this section. Discuss any difficulties and provide further examples if necessary. Ask the learners to do the exercises. Provide assistance if necessary. Check the answers orally.

Speaking

There are three activities in this section. The first two (**Tasks 11** and **12**) are recorded and the student is required to pause the recording and make appropriate responses. **Task 13** (role play) enables pairs of learners to simulate real calls and apply the language they have learnt in the course of the unit.

In the first Speaking exercise (**Task 11**), introduce the language point that is exercised. Give further examples if necessary and then allow students to listen to the cassette first to help them if necessary. Ask the class to listen to the prompts on the recording and produce appropriate responses before they hear the model responses. This work can be done by the whole class, groups, pairs or individuals. **Task 12** is best done individually. Help to set the scene and allow students to listen to the recording first if necessary. Ask students to repeat the exercise for further practice. If you have access to a language laboratory, it could be of use when doing these exercises.
The role plays (**Task 13**) can be done by learners in pairs. In each of the role plays notes are provided for the caller (A) and the person who answers (B). The A notes are included in the units while the B notes are at the back of the book (pages 72–76). Each pair of learners can do any or all of the role plays in order. They may also reverse roles after the first completion of the role play. The role plays should not be attempted until you are reasonably confident that your learners have mastered the relevant language since this activity is an opportunity for free application and expression and is not easily monitored.

Overview

	Listening	Language study	Speaking
Unit 1	Identifying people	Requesting information Countries and nationalities	Spelling Role plays
Unit 2	Connecting people Wrong numbers	Asking questions Note-taking	Giving dates Role plays
Unit 3	Enquiries for prices and discounts	Passing on messages Note-taking	Abbreviations/spelling Role plays
Unit 4	Ordering	Talking about the future Nouns and verbs	Giving references and numbers Role plays
Unit 5	Hotel and travel arrangements	Probability and possibility Reporting questions	Question tags Role plays
Unit 6	Changing arrangements	Future possibilities Phrasal verbs	Giving information/spelling Role plays
Unit 7	Checking up on problems	Apologising Getting things done	Figures and calculations Role plays
Unit 8	Making and handling complaints	Fault diagnosis Nouns and verbs	Giving information/spelling Role plays

1 How can I help you?

Listening

Task 1

Listen to two phone conversations and complete the table.

Call	Where is the person the caller wants?	Country of meeting	Number of caller
1			
2			

Task 2

Listen to the two calls in Task 1 again. Write a message for each of the people the callers wanted.

1

Phone Message

Caller: David Bartlett ……………………

Message: ……………………

……………………

……………………

……………………

……………………

……………………

Anne

2

Telephone Message

Message for: ……………………

Call from: ……………………

Message: ……………………

……………………

……………………

……………………

……………………

……………………

You will find the tapescript on page 77.

What to say – what to expect

Read these useful sentences and make sure you understand them. Use a dictionary to help you if necessary.

Announcing identity

Person calling
Hello, this is Barbara Ling.
My name's Daniel Wong.
Good morning. It's Rebecca Park here.

Person called
Hello. David Jackson.
Can I help you?
Marketing Services PLC. Good afternoon.
Supersport.com

Asking if someone is in

Person calling
Can I speak to Mr Lee, please?
Hello, is Laura there?
Hi, it's Jim Wilson here. Is Sandra in?
Could you put me through to Maria Garcia, please?
Can I speak to someone in Marketing, please?

Person called
Hold the line, please.
Hold on, please, I'll see.
Yes, I'll just get her.
I'll just transfer you.
Yes, just a moment.

Person wanted is not there

Person called
I'm afraid she isn't in at the moment.
Sorry, he's just gone out. Would you like to call back later?
She's away for a few days. Can I give her a message?
He's out of the office this week, I'm afraid.
You can contact her on her mobile. The number is 07700 900008.

When will the person wanted be in?

Person calling
What time will she be back?
Will he be back later today?
Can I contact her tomorrow?
When would be a good time to call again?

Person called
She should be back by 4 o'clock.
We're expecting him at around 11.00.
She's due back tomorrow.
Why don't you try in a couple of hours?

Calling off

Person calling
I'll get back to you soon.
Thanks very much. Goodbye.
OK. Bye.

Person called
Thanks for calling.
We'll be in touch about it soon. Goodbye.
Bye.

Task 3

Complete the sentences with words from the list below. Use each word once only.

1 Hello, is that Mauro ....?

2 Just a .... , please.

3 Wait a minute, I'll .... if she's here.

4 I'll get the information you want. Do you mind .... on?

5 You should be able to reach her on her ....

6 Try calling back .... an hour's time.

7 Sorry, he's not .... at the moment.

8 I'll have to put you on .... while I check.

here	in	mobile	speaking	hold
see	moment	holding		

Task 4

Choose the best responses.

1 I'd like to speak to Ms Chan, please.
a Yes.
b I'm afraid she's not here at the moment.
c Well, you can't.

2 Can I speak to Mr Ramirez, please?
a Hold on, please.
b Don't go away.
c All right.

3 Who's speaking?
a I am called Pierre Marceau.
b My name's Pierre Marceau.
c Pierre Marceau is speaking.

4 Could I speak to Marta Owen, please?
a Who's calling?
b Who are you?
c What's your name?

5 Can I call you back later?
a Yes, call me.
b Yes, please do.
c Of course call, yes.

6 When will she be back?
a One hour.
b After one hour.
c In an hour's time.

Task 5

Listen to the phone conversation once and decide which message pad has the correct information.

1

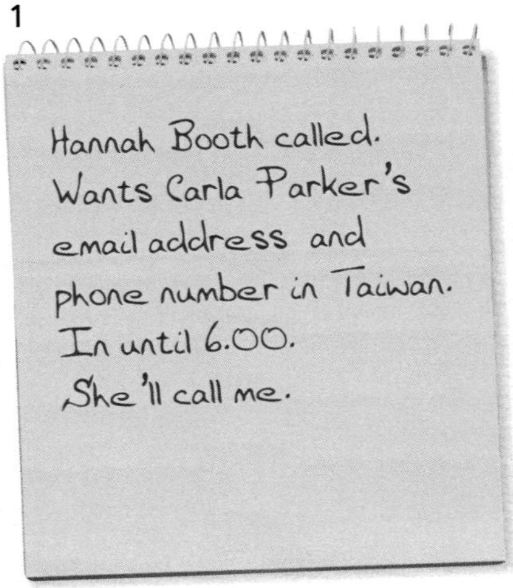
Hannah Booth called.
Wants Carla Parker's
email address and
phone number in Taiwan.
In until 6.00.
She'll call me.

2

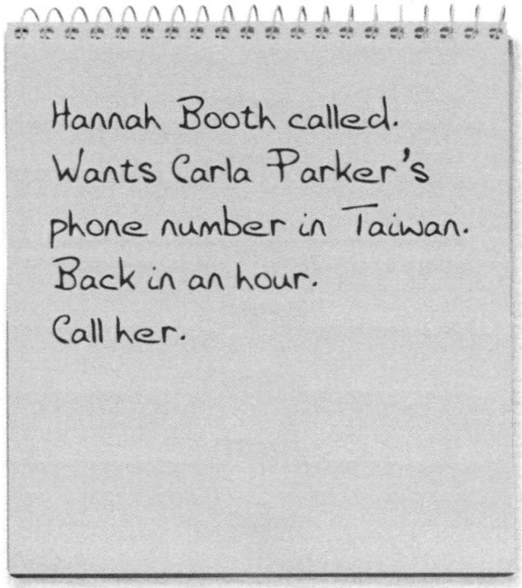
Hannah Booth called.
Wants Carla Parker's
phone number in Taiwan.
Back in an hour.
Call her.

3

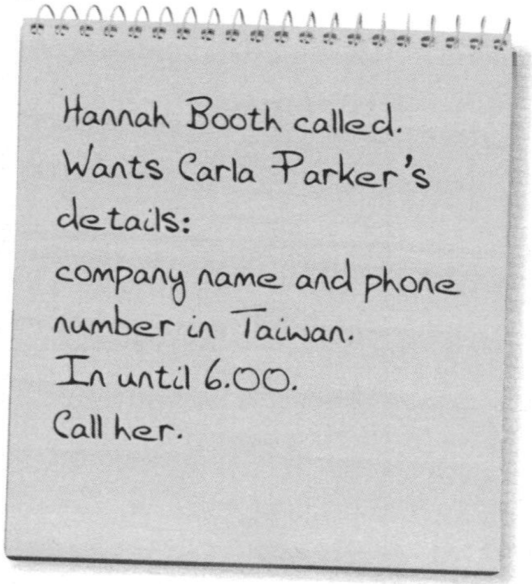
Hannah Booth called.
Wants Carla Parker's
details:
company name and phone
number in Taiwan.
In until 6.00.
Call her.

Listen again and complete the table. Then answer the questions.

Person called	Caller	Request	Who will make the next call?

1 What sort of work does Carla Parker do?
2 How is Richard Dawson going to find out the information?

You will find the tapescript on page 78.

Task 6

Listen to the phone conversation and complete the table. Then listen again and answer the questions.

Person called	Caller	Request	Who will make the next call?

1 What have Star Cars International ordered from Motor Systems UK?
2 What's the order number?
3 When would Star Cars International like delivery of their order?

You will find the tapescript on page 78.

What to say – what to expect

Read these useful sentences and make sure you understand them. Use a dictionary to help you.

Requests

Person calling

I'd like to speak to somebody about …

Can you give me some information about … ?

What's the position on … ?

We'd like an earlier delivery date if possible.

Could you bring delivery forward by a few weeks?

Person called

What's the order number?

Can you give me the reference number?

When did you send the order?

I'll have to check with the department concerned.

I can't tell you right now, but I can look into it and get back to you.

Can I let you know the situation tomorrow?

Task 7

Listen to Richard Dawson and Mark Wheeler phoning back, as they said they would. Write notes about the two calls on the message pads.

1

2

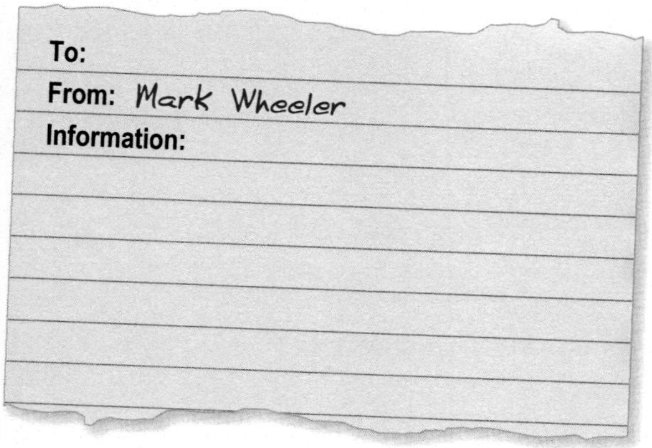

You will find the tapescript on page 79.

Task 8

Complete these two conversations with sentences from the list below. Use each sentence once only.

A: Hello, is that Motor Systems UK?

B: **1** ....

A: Can I speak to Mark Wheeler, please?

B: **2** ....

A: OK. Do you know what time he will be free?

B: **3** ....

A: Right, I'll call again then. Thanks very much.

B: **4** ....

A: Goodbye.

C: **5** ....

D: I'd like to speak to someone about bringing forward a delivery date.

C: **6** ....

E: **7** ....

D: I'm phoning about our order for some special plugs.

E: **8** ....

D: Yes, it's MS/72/03. We'd like an earlier delivery date if possible.

E: **9** ....

D: OK. Could you call me back today?

E: **10** ....

D: That'll be fine. Thanks very much.

a I'll put you through to Order Enquiries.
b From about three this afternoon.
c Yes, later this afternoon if that's convenient.
d Yes, it is. Can I help you?
e I'm afraid he's in a meeting at the moment.
f Motor Systems UK. Can I help you?
g Right. Well, I'll have to check with the factory supervisor.
h Can you give me the order number?
i Order Enquiries. Can I help you?
j You're welcome. Goodbye.

Language study

Task 9 Requesting information

Study these examples of how to ask for information politely.

You don't know a caller's name. (give)
Could you give me your name, please?
You aren't sure of the name of the caller's company. (repeat)
Would you repeat the name of your company, please?
You want to know where the caller is calling from. (tell)
Can you tell me where you're calling from, please?

Could and *would* are more polite than *can*.

Now make questions using *could*, *would* and *can* in a similar way.

1 You aren't sure exactly what the caller is phoning about. (tell)
2 You want to know the caller's telephone number. (give)
3 You don't know how to spell the caller's name. (spell)
4 You didn't hear the caller's address clearly. (repeat)
5 You want to find out when the caller will be in the office tomorrow. (tell)
6 You aren't sure about the delivery date of your order. (confirm)

Task 10 Countries and nationalities

Complete the table with the missing countries and nationalities. Use a dictionary to help you if necessary.

	Nationality	Country		Nationality	Country
1	....	China	9	Swiss	....
2	American	....	10	....	Brazil
3	Korean	....	11	Taiwanese	....
4	....	France	12	....	Sweden
5	German	....	13	....	UK
6	....	Japan	14	Belgian	....
7	Spanish	....	15	....	Saudi Arabia
8	....	The Netherlands	16	Irish	....

Speaking

Task 11

Listen to the callers who ask you how to spell these names.
Pause the recording after each caller and spell the names. Then listen to the correct spelling. You may listen to the recording first to help you.

1 Wallace	**2** Lefevre	**3** Schoppen
4 McDonagh	**5** Takamura	**6** Cricchi

You will find the tapescript on page 79.

Task 12

1 You work in an office with Julia, Fernando and Kirsten. Look at the schedule, which shows where your colleagues will be during the day. Listen to the callers who want to speak to your colleagues. Pause the recording after each caller and respond. Do the task twice. The first time it is 11.30 a.m. The second time it is 3 p.m. You may listen to the recording first to help you.

SCHEDULE Wednesday 5 May

	Julia	Fernando	Kirsten
9–10	Will be in late – have to take car to garage	Visiting a new client – contact me on mobile if urgent. Should be back in office by 12.00	Plan to be in office early, preparing for meetings
10–11			
11–12	Sales meeting part 1– do not disturb		Visiting International Computers – meeting and lunch there – contact on mobile if urgent
12–1			
1–2	Lunch with Sales team		
2–3	Sales meeting part 2 – should finish at 3.30 – do not disturb	Working at home all afternoon – ring me there	Back in office
3–4			Meeting Human Resources Director and team in meeting room upstairs – do not disturb
4–5	Meeting with design team – at least two hours – contact on mobile		
5–6			Must leave office at 5.00 on the dot!
	Home: 5784881 Mobile: 07890 376291	Home: 6740035 Mobile: 07773 925586	Home: 3954936 Mobile: 07966 484912

You will find the tapescript on page 80.

2 Now you are making the following calls. Listen to the person who answers your call. Pause the recording and respond. You may listen to the recording first to help you.

a You have just phoned your colleague, Kirsten, as you need to give her a long message.
b You want to speak to Mr Wheeler at Motor Systems UK.
c You want to ask Mr Wheeler about the price of QP pump motors.
d You've just asked to speak to Carla Parker at Atlas Imports and Exports. If she's not in, leave a message for her to call you back. Leave your phone number.

You will find the tapescript on page 80.

Task 13 Role play

Work with another student when you do this exercise. Agree which of you is Student A and which is Student B. Student A has information on this page, Student B on page 72.
Sit back to back. Student A should now 'call' Student B. When you have done the calls once, change roles.

A1 You are a colleague of Hannah Booth. You would like to ask Richard Dawson for the company name and phone number of someone called Kevin Kim in South Korea who he mentioned to Hannah Booth.

A2 You are a colleague of Nick Sheridan. You would like to ask Mark Wheeler the price of plugs (reference number MS/74/07) from Motor Systems UK.

A3 You are a colleague of Carla Parker. Carla has asked you to call Richard Dawson as she would like some information about Hannah Booth's company. If he is unavailable, leave a message, and explain that Carla would like the information urgently.

2 Hold the line, please

Listening

Task 1

Listen to two phone conversations and complete the table.

Call	Caller's name	Person wanted	Person answering
1			receptionist
2			

Task 2

Listen to the two calls in Task 1 again. Decide if the statements about the calls are true (T) or false (F).

1 Renata Schatke tried to make a call straight to Jim Channon. **T/F**
2 The receptionist sent someone to look for Mr Channon. **T/F**
3 The callers in both calls were asked to wait. **T/F**
4 Liz Hunt did not answer the phone at her own desk. **T/F**

You will find the tapescript on page 80.

What to say – what to expect

Read these useful sentences and make sure you understand them. Use a dictionary to help you if necessary.

Connecting to an extension

Person calling

Could you put me through to Lorenzo Rinelli, please?
I'd like to speak to Isabel Silva, please.
Could you give me the number of his direct line, please?
She asked me to phone her this morning.

Person called

Would you hold on, please?
Please hold the line.
Yes, please hold on and I'll put you through.
Sorry to keep you waiting.
Sorry, this isn't her extension. I'll try to transfer you.

Wrong number

Person calling

Oh, isn't that International Computers?
I'm sorry, I must have the wrong number.
I thought I'd dialled 01632 875 4903.
Well, I found this number on a price list.
Sorry. I must have written down the wrong number. I'll try Directory Enquiries.
Sorry to have bothered you.
I must have got the area code wrong.

Person called

I think you must have dialled the wrong number. What number have you got?
I'm afraid there's nobody here with that name.
Sorry, they moved last August. The new number is 0117 496 0003.
You could probably find the right number on the Internet.

Task 3

Complete the sentences with words from the list below. Use each word once only.

1 Isn't that 423884? I think that's what I ....

2 No, this isn't her .... I'll transfer you back to the switchboard.

3 I'm sorry to have .... you.

4 He's out of the office at the moment. You'll get him if you call his ....

5 I've lost their new number. I'll have to call .... Enquiries.

6 You might find what you want if you look on the ....

7 Sorry to keep you waiting. Please .... on a bit longer while I try to find her.

8 Could you please tell me the area .... for Hamburg?

9 I'm now in a position to .... the arrangements we made.

10 Can we reschedule the .... for next week?

hold	Directory	confirm	Internet	appointment
extension	code	mobile	dialled	bothered

Task 4

Choose the best responses.

1 Can you put me through to Georg Stiess, please?
  a I'll see if he's in his office at the moment.
  b I've got the wrong number.
  c I'll check again.

2 Isn't that Seattle, then?
  a No, the number has changed.
  b No, you must have the wrong area code.
  c Sorry, I may have dialled the wrong extension.

3 You asked me to confirm their email address.
  a Yes, that's the most likely one.
  b Yes, let me just write it down.
  c Yes, let me find out.

4 No, this isn't an insurance company.
- a I'm sorry to have bothered you.
- b I'll call again later.
- c Can you put me through to Galina Klimov, please?

5 Mr Chan asked me to call this morning.
- a Sorry, you've got the wrong name.
- b Do you know the area code?
- c I'm afraid there's nobody with that name here.

6 We can let you know what the alternatives are.
- a Thank you. I know them.
- b Thanks. I'm glad that's OK now.
- c Thanks. I can order what we need then.

Task 5

Listen to the phone conversation and complete the table.

Person wanted	Company wanted	Company answering

You will find the tapescript on page 81.

What to say – what to expect

Read these useful sentences and make sure you understand them. Use a dictionary to help you if necessary.

Making and confirming arrangements

Person calling / Person called
I'll email you the details.
I'll fax you to confirm all the arrangements.
Can I get back to you to confirm those details?
Right, I've got the details you were asking about.
You asked me to call back to let you know if the terms were acceptable.
We've discussed possible dates for the meeting, and next Thursday would suit us all.
Could you manage that?
I'll put everything we've discussed in writing and copy it to the relevant people.
About the delivery times – the earliest date we can manage is 1st May.
When you've checked things at your end, could you let me know, please?
There will be a car to collect you from the airport when you arrive.

Task 6

Listen to the phone conversation and complete the table.

Caller's name	Person called	Reason for calling

Listen again and answer the questions.

1 Why is the consignment a little delayed?
2 How many containers are in the shipment?
3 Why will Frank Patterson probably not visit Italy in this half of the year?

You will find the tapescript on page 81.

Task 7

You would like to renew the insurance on your car. You ring the insurance company. Listen to the choices on the recorded menu and tick the appropriate box in order to speak to someone in the right department.

1 ☐ 2 ☐ 3 ☐ 4 ☐

Listen again and answer the questions.

1 What number do you press if you want to ask about payment?
2 What number do you press if you want to make a claim?
3 If you have a different query, what should you do?

You will find the tapescript on page 82.

Task 8

Complete these two conversations with sentences from the list below. Use each sentence once only.

A: **1** ....

B: Hello. Could I speak to Joelle Duval, please?

A: **2** ....

B: Linda Topley.

A: **3** ....

B: She said she'd be in all morning.

A: **4** ....

C: Joelle Duval speaking.

B: **5** ....

C: Oh, yes, thank you, it's about …

D: Carl Anderson.

E: **6** ....

D: Lindberg, did you say?

E: **7** ....

D: There's no one here with that name.

E: **8** ....

D: Well, this is 897413.

E: **9** ....

D: That's all right. Goodbye.

a Yes, that's right.
b You asked me to call as soon as possible.
c Hold the line, please, and I'll see if she's there.
d Who's calling, please?
e Oh, isn't there? I thought I'd dialled 897412.
f Could I speak to Ingrid Lindberg, please?
g Right, I can connect you now.
h International Recruitment Services, good morning.
i Oh, I'm very sorry. I must have dialled the wrong number.

Language study

Task 9 Asking questions

Study these examples of how to make questions.

You want to know where the nearest payphone is.
Where's the nearest payphone?
Find out how to spell someone's name.
How do you spell your name?

Now make questions in a similar way.

1 Find out when Ms Gonzalez will be back.
2 You'd like to know why the sales office hasn't called.
3 Find out when the manager normally arrives at the office.
4 You want to know why the documents have been delayed.
5 Find out the number to dial for Directory Enquiries.
6 You'd like to know where someone is phoning from.
7 You need to know a convenient time to ring someone back.
8 Find out why the meeting has been postponed and what the new date is.

Task 10 Note-taking (1)

It is often necessary to take notes during phone conversations. You can do this by shortening words, like days of the week, and using abbreviations. The notes may not be just for your own use, but other people may need to be able to understand them too. The following are often useful:

Re/re = about, regarding, on the subject of
e.g. (Latin: exempli gratia) = for example
NB (Latin: nota bene) = note, notice especially
i.e. (Latin: id est) = that is, like
p.a. (Latin: per annum) = per/each year
c = about, approximately
cf. = compare with
a.m. = morning
p.m. = afternoon
eve = evening
HQ = headquarters
MD = Managing Director
info = information
asap = as soon as possible
sb = somebody/someone
@ = at
v (versus) = against
& = and
pl/pls = please
no. = number

Rewrite these notes in full sentences.

1 Meet Rosalia Tues @ 3 p.m.
2 Send Mauro info re sales figs asap.
3 Can sb go to HQ on Thurs a.m?
4 Urgent meeting Mon a.m. re sales & stock figs.
5 Ring Gina asap – NB out after 2.

Now put these sentences into note form.

6 Please call Adriano about the meeting on Tuesday afternoon.
7 Headquarters want some information about the number used each year.
8 Hiromi needs figures for the presentation, for example, the budget figures compared with the sales figures for September.
9 There will probably be about 200 people at the conference in February.
10 Note that the Managing Director will be away from Tuesday to Thursday.

Speaking

Task 11

Listen to the callers who ask when certain things happened or will happen. Pause the recording and tell them, using the dates given. Then listen to the correct way to say them. You may listen to the recording first to help you.

1 9 July 2004
2 17/9/01 (BrE)
3 Wednesday June 12
4 7 December 1983
5 Tuesday 25 April
6 2/11/03 (AmE)
7 Thursday March 15
8 29 August 1999
9 10 May 2006
10 21/10/12 (BrE)

In British English, dates using only numbers give the day, then the month, then the year, e.g. 8 May 2002 = 8/5/02

In American English, the order is month, day, year, e.g. 8 May 2002 = 5/8/02

You will find the tapescript on page 82.

Task 12

You are making the following calls. Listen to the person speaking. Pause the recording and respond.
You may listen to the recording first to help you.

1 You are phoning Paperworks.
2 You want to speak to Liz Hunt because she asked you to call today. If she's not there, leave a message to say you called.
3 You are calling Frank Patterson to confirm an appointment you've been trying to arrange. You suggest Monday.
4 You work in the Public Affairs department. You have just picked up your phone.

You will find the tapescript on page 82.

Task 13 Role play

Work with another student when you do this exercise. Agree which of you is Student A and which is Student B. Student A has information on this page, Student B on page 72.

Sit back to back. Student A should now 'call' Student B. When you have done the calls once, change roles.

A1 You are a colleague of Liz Hunt at Orbis Group. Call Yoshida Tokuko in his office in Tokyo, on 0081 3 3486 5912. You would like him to confirm he can meet Liz Hunt when she comes to Tokyo on 14 May.

A2 Try the same call again.

A3 You are a colleague of Frank Patterson. He has asked you to call Teresa Lombardo urgently because the consignment sent by her company hasn't arrived yet – she had said it would arrive by the end of May and it's now 7 June. Find out why it is late and when it is expected to arrive.

3 Making enquiries

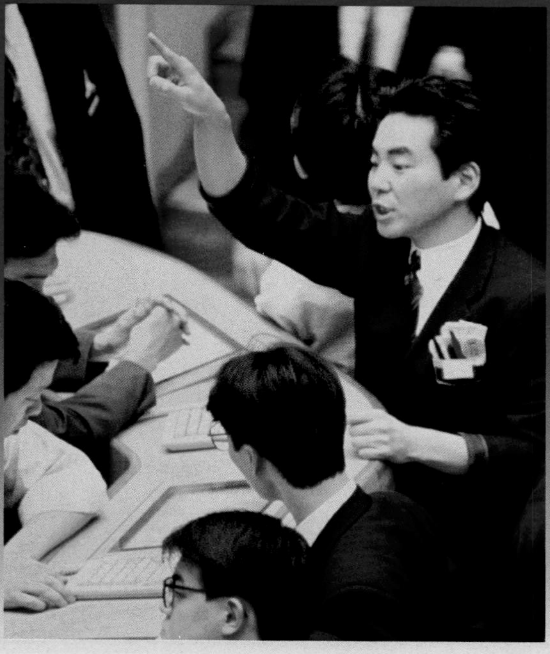

Listening

Task 1

Listen to two phone conversations and complete the table.

Call	Caller	Person/Company called	Caller interested in
1			
2			

Task 2

Listen to the calls in Task 1 again. Take notes on the message pads.

1

2

You will find the tapescript on page 83.

What to say – what to expect

Read these useful sentences and make sure you understand them. Use a dictionary to help you if necessary.

Enquiries for prices and discounts

Person calling

I've seen your advertisement and I'd like to know how much you charge for …

Could you please tell me what your terms are?

Does the discount go up according to the size of the order?

What are your hotel rates? Does that include breakfast?

Person called

We can give you a 10% discount if you order before 30 October.

The sale is going to continue for the next month.

When you open an account with us you get 5% off immediately.

Our prices start at $100 for a single room, with breakfast included.

Task 3

Complete the sentences with words or phrases from the list below. Use each word or phrase once only.

1. Here are the .... prices available for the flights you wanted.
2. Couldn't you manage to .... me a better discount for this large .... ?
3. It's a good price – it hasn't .... since last year.
4. We've had to .... prices in line with inflation.
5. We can offer you a .... discount if you order by the end of the month.
6. Our room .... compare favourably with similar hotels in the area.
7. How much do you .... for each transaction?
8. I think investing in .... in the dotcom sector may be risky now.

increase	rates	order	gone up	charge
shares	lowest	give	special	

Task 4

Choose the best responses.

1 Can you give me a quote?
- a We haven't any more available.
- b This price is very competitive.
- c It will be $350.

2 Can we have a higher discount?
- a It depends on the number you order.
- b The prices are our lowest.
- c It's not so much.

3 We are thinking of buying your products.
- a Then take advantage of our introductory offer.
- b Business is good at present.
- c Share prices have been falling lately.

4 Can you offer the large size at the same price?
- a No, it's cheaper.
- b No, it's more expensive.
- c No, the price is unchanged.

5 When do we need to pay the balance?
- a Please pay by bank transfer.
- b No credit is allowed.
- c By 30 May.

6 Are those the best prices you can offer?
- a Yes, we have plenty available.
- b Yes, they are fixed for six months.
- c Yes, they are very important.

Task 5

You would like to book some tickets for the cinema, but you want some information first. Listen to the recorded menu and complete the notes on the message pad.

To book tickets: ....

To find directions to the cinema: ....

Prices:

Adult: Standard .... Superior ....

Students / Senior citizens: Standard .... Superior ....

Children under ....: Standard .... Superior ....

Family ticket: ....

You will find the tapescript on page 84.

What to say – what to expect

Read these useful sentences and make sure you understand them. Use a dictionary to help you if necessary.

Enquiries for prices and discounts

Person calling

Can I order online? What's your website address?
I've got your February price list. Is it still valid?
We usually get a better discount on a repeat order.
As this is such a major order, we expected a better discount.
Why have you reduced the discount?
Can you quote me a price for that?

Person called

I can email our price list to you, or shall I fax it?
Would you like our special introductory offer?
Those are the best terms we can offer.
You'll find our prices can't be matched.
The price includes insurance and delivery by courier.
You'll find all our prices and terms on the website.

Task 6

Listen to the phone conversation and complete the email.

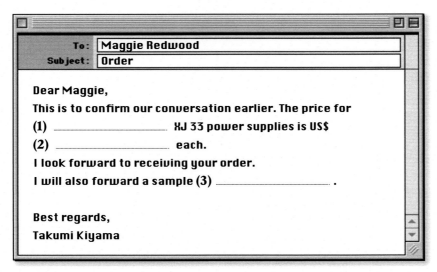

To: Maggie Redwood
Subject: Order

Dear Maggie,
This is to confirm our conversation earlier. The price for
(1) .... XJ 33 power supplies is US$
(2) .... each.
I look forward to receiving your order.
I will also forward a sample (3) ....

Best regards,
Takumi Kiyama

Listen again and answer the questions.

4 Which power supply model has Maggie Redwood's company been using up to now?
5 Why is Takomi Kiyama going to send an XJ 44M to Maggie Redwood?

You will find the tapescript on page 84.

Task 7

Listen to the phone conversation and complete the notes.

Conference Centre wants: Bebbington Porcelain Blue Room Collection
- tableware service for **(1)** .... *people.*
Sales Director offers: special discount of **(2)** ....
Normal discount is **(3)** ....
Complete boxed tableware set costs **(4)** ....
Are pasta dishes and dessert bowls included? **(5)** ....
Same terms for follow-up orders? **(6)** ....

Listen again and answer the questions.

7 What is Eva Frei going to do next, following the phone call?
8 How soon does she think the conference centre may place the order?

You will find the tapescript on page 85.

Task 8

Complete the conversation with sentences from the list below. Use each sentence once only.

A: PC Delivery. Good morning.

B: Could I speak to Anita McGarry, please?

A: **1** ....

B: I'm calling from Computer Sales Ltd. We'd like to order some DE960 printers.

A: **2** ....

B: **3** ....

A: Oh, yes, until the end of the year.

B: **4** ....

A: You've done business with us before, haven't you?

B: Yes, and this is our second order for this type of printer.

A: **5** ....

B: **6** ....

A: Oh, we don't normally go over 10%.

B: **7** ....

A: I see. Well, I'd better confirm that with someone in my department.

B: **8** ....

A: Yes, Computer Sales Ltd, you said. And your name is ... ?

B: John Draper.

A: **9** ....

a I see. How many would you like?
b We're thinking in terms of 12%. How does that sound?
c Yes, please do that, and then perhaps you'll call me back.
d Right, Mr Draper. I'll call you back later this morning.
e But we had 7% last time, and we were told it would be 5% higher for a repeat order.
f What discount would you offer on an order for 100?
g Speaking.
h That's good. We give a higher discount on a repeat order.
i Well, it depends on your terms. Is your May price list still valid?

Language study

Task 9 Passing on messages

Study these examples of how to pass on messages.

'I'm arriving on flight BA 532,' said Claudia Peuser.
Claudia Peuser said she was arriving on flight BA 532.
Ming Li said to Jeff Shen, 'Please send confirmation in writing to the suppliers.'
Ming Li told/asked Jeff Shen to send confirmation in writing to the suppliers.

Now pass on these messages in a similar way. Make sure that you make all the necessary changes.

1 'I'll give you an extra 2% discount for such a big order,' Prisca Marchal said to me.
2 'Alicia, please tell Pablo Lubertino we've received his order,' the manager said.
3 'How do you spell your second name?' the receptionist asked Xin Yuzhuo.
4 The Sales Manager said to me, 'Tell her we'll offer them a bigger discount.'
5 Mete Irmak said, 'We paid the account by bank transfer on 17 October.'
6 'Can you check whether the figures in the file are correct, please?' Daniel Tai said to Hanna Chang.
7 'Could you tell Abdullah Hassan that I called, please?' Melissa Fu said to the receptionist.
8 Kenny Liu said to his colleague, 'Is the sale due to end next week?'

Task 10 Note-taking (2)

Choose the abbreviation from the list below that matches each of these words and phrases.

1 Personal Assistant
2 and so on
3 maximum
4 stamped addressed envelope
5 as soon as possible
6 Research and Development
7 per annum/year
8 for the attention of
9 cost, insurance, freight
10 note
11 estimated time of arrival
12 Managing Director
13 about, on the subject of
14 for example
15 thousand
16 especially
17 Greenwich Mean Time
18 information

ETA	re	e.g.	NB	Attn	CIF	asap	esp	SAE
etc.	GMT	MD	K	max	R&D	info	PA	p.a.

Now use abbreviations to put these sentences into note form.

19 Could you ask Tatiana about the invoice as soon as you can?
20 The cost will be $49,000 including insurance and freight.
21 The interest payable will be 18 per cent per year.
22 It is very important that we don't pay more than €1,250.
23 The Managing Director is expected to arrive at half past three on Wednesday afternoon.
24 Please send each applicant a stamped addressed envelope.

Speaking

Task 11

Listen to the callers. Pause the recording and answer their questions, using the information given.
You may listen to the recording first to help you.

1 Tiphaigne
2 ETA 10.25 a.m.
3 Keumsung
4 2.30 p.m. on 17 July
5 €130
6 £3.75
7 asap
8 Form + SAE

You will find the tapescript on page 86.

Task 12

Read this email from a friend in Madrid.

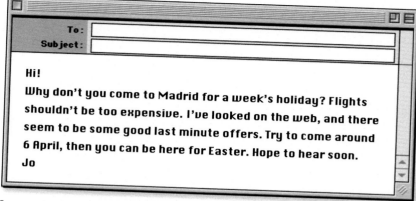

> To:
> Subject:
>
> Hi!
> Why don't you come to Madrid for a week's holiday? Flights shouldn't be too expensive. I've looked on the web, and there seem to be some good last minute offers. Try to come around 6 April, then you can be here for Easter. Hope to hear soon.
> Jo

You have looked at different websites for information about flights, but the fares for the Easter period look rather high. You can't find the last minute offers your friend mentioned. You call a travel agent to see if they can help. You don't want to pay more than €200. You don't mind what time the flights are. Listen to what the travel agent says. Pause the recording after each question and respond.
You may listen to the recording first to help you.

You will find the tapescript on page 86.

Task 13 Role play

Work with another student when you do this exercise. Agree which of you is Student A and which is Student B. Student A has information on this page, Student B on pages 72–73.
Sit back to back. Student A should now 'call' Student B. When you have done the calls once, change roles.

A1 You have inherited some money and you would like to invest £10,000 in shares. You call Daniel Evans at Capital Investment Services. You want to invest in 'safe' companies and you would like some prices, and information on how you can expect your money to grow.

A2 Your office is in London. You have to go on a business trip to Lima, Peru, at short notice for a ten-day trip. You call Globe Travel Agency to ask Annabel Davies to make bookings for you. You would like to stop over in Mexico City on the way, and in Port of Spain, Trinidad on the return trip. You would like to travel business class. Tell the travel agent the dates you need to travel. You need to know how much the return fare will be before you can make the booking.

A3 You are a colleague of Ben Corbett, the Sales Director at Bebbington Porcelain. Ben has asked you to call Eva Frei at the International Shop in Berlin to give her some more information about the tableware which the conference centre may order. There is a three-month delay in the production of the Blue Room Collection dinner plates, and stocks are low now; the Violet Room Collection is very similar, and all the pieces are in stock; the Gold Room Collection tableware has been discounted by a further 5% for three months.

4 Placing an order

Listening

Task 1

Listen to three phone calls and complete the table.

Call	Company called	Caller	Reason for calling
1			
2			
3			

Task 2

Listen to the calls in Task 1 again. Decide if the statements about the calls are true (T) or false (F).

1 Ms Penella hasn't got her luggage with her. **T/F**
2 She will wait for the taxi inside the restaurant. **T/F**

3 The mail order company already knows the customer's address. **T/F**
4 Post and packing are not included in the price. **T/F**

5 If you want to order the latest brochure you have to press 2. **T/F**
6 When you have completed making the order you have to press 5. **T/F**

You will find the tapescript on page 86.

What to say – what to expect

Read these useful sentences and make sure you understand them. Use a dictionary to help you if necessary.

Ordering

Person calling
We're ready to order now.
We'd like to place an order for 200 packs of printer paper.
I'm phoning you with a repeat order.
Have you got everything in stock?
It's a very urgent order.
When will you get more stock in?

Person called
What would you like to order?
Can I have your customer reference number?
What's the item number in the catalogue?
I'm afraid that item is out of stock.
The order will be processed this week.
New stock is due in next month.
We've got a backlog, so the order won't be dispatched for at least two weeks, I'm afraid.

Delivery

Person calling
We need the goods urgently. Can you dispatch them today?
How soon will the parcel arrive?
How will the order be sent?
When can we expect delivery?

Person called
We'll dispatch the goods immediately from stock.
We'll send the goods by next day courier.
The consignment will be sent by air freight.
It should arrive by the end of the week.

Avoiding misunderstandings

Person calling / Person called
Sorry, I didn't hear what you said. Could you repeat the price, please?
I didn't catch what you said. Could you repeat it more slowly?
It's a very bad line, I'm afraid. Can I call you back?
I'm on my mobile and the line's not very good.
I think the signal is going. I'll call back when I can.
I can hardly hear you. Can you speak up, please?

Task 3

Complete the sentences with words from the list below. Use each word once only.

1 Thank you for your quotation. We'd like to .... an order now.

2 I need to .... down the reference number.

3 When you order, you need to give the .... number from the latest ....

4 We're under pressure from our customer. Can you send the order .... , please?

5 I don't need to .... by credit card because I've got a monthly ....

6 The .... will be enclosed with the goods.

7 I'm afraid it's out of ...., so it won't be sent for two weeks.

8 We'll send it by air ...., so it will arrive tomorrow.

9 I'm calling to make a .... order. We'd like exactly the same as last month.

10 Sorry, I didn't quite .... what you said. Could you say that again?

freight	invoice	repeat	pay	stock	item
urgently	note	catch	account	place	catalogue

Task 4

Choose the best responses.

1 Where are you calling from?
a I'm on the phone.
b Madrid.
c This is Dolores speaking.

2 Could you quote the reference number, please?
a I've received your quote.
b Yes, that's the number I want.
c Hold on a moment, I'll just find it.

3 How would you like to pay?
a By credit card.
b I'll pay the invoice next week.
c I'll send you a quote.

4 When will the order be dispatched?
a Last Tuesday.
b Within three working days.
c It's in stock now.

5 Can I call you back in five minutes?
a Yes, of course.
b What's your mobile number?
c The signal's gone very faint.

6 We've got some more on order.
a So you'll have to order some more?
b We'd like to order some more.
c When will they be in stock?

Task 5

Listen to the phone conversation and decide which order confirmation form has the correct information.

1

Blooming Flowers

Order Confirmation
Order placed by: James Elliott
Type of flowers: Roses – red
Quantity: 50
Person addressed to: Caterina Santiago
Delivery address:
43 Pennsylvania Avenue, Bloomington
Delivery date/time: 6 May, 12 p.m.
Message:
Happy Birthday, All my love, J
Payment: Cheque ☐ Credit card ☑

2

Blooming Flowers

Order Confirmation
Order placed by: James Elliott
Type of flowers: Roses – red
Quantity: 15
Person addressed to: Caterina Santiago
Delivery address:
43 Pennsylvania Avenue, Bloomington
Delivery date/time: 12 May, 6 p.m.
Message:
Happy Birthday, All my love, J
Payment: Cheque ☐ Credit card ☑

Listen again and decide if the statements about the call are true (T) or false (F).

1. James Elliott had decided before he made the call what he wanted to order. **T/F**
2. He asks for the flowers to be sent to his home address. **T/F**
3. The florist checks the number of flowers he wants to order. **T/F**

You will find the tapescript on page 88.

What to say – what to expect

Read these useful sentences and make sure you understand them. Use a dictionary to help you if necessary.

Confirming order arrangements

Person calling / Person called
Can we go through the order in detail, please?
Can I double-check some of the order details?
We've got a problem with some of the items on the order.
I'm calling to confirm the arrangements we agreed.
I'll email you confirmation of everything we've discussed.

Task 6

Listen to the phone conversation and complete the email.

To: Serge Duval
Subject: Order confirmation

Dear Serge,
This is to confirm our conversation earlier.
You have ordered (1) .... CM 25 hard drives at a reduced price of (2) .... each rather than (3) .... because the order is significantly larger.
You will settle the account by (4) .... transfer immediately, to the amount of $132,000.
Thank you for your order.
Jennifer Sato
Kobayashi Components

Listen again and answer the questions.

5 What quantity was Serge Duval originally interested in?
6 What is included in the price?

You will find the tapescript on page 88.

Task 7

Listen to the phone conversation and complete the order.

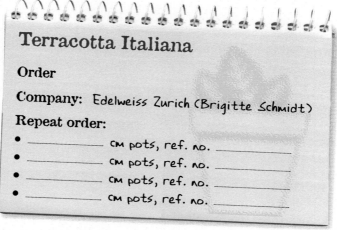

Terracotta Italiana
Order
Company: Edelweiss Zurich (Brigitte Schmidt)
Repeat order:
- .... cm pots, ref. no. ....
- .... cm pots, ref. no. ....
- .... cm pots, ref. no. ....
- .... cm pots, ref. no. ....

Listen again and answer the questions.

1 What is the item number of the pot Brigitte is enquiring about?
2 What does Brigitte want to know it is guaranteed against?
3 How soon will Alessandra be able to answer Brigitte's question?
4 When will the order be dispatched?
5 How will the order be delivered?

You will find the tapescript on page 89.

Task 8

Complete the conversation with sentences from the list below. Use each sentence once only.

A: Expedia, Sara speaking. Can I help you?

B: **1** ....

A: Certainly. Have you ordered with us before?

B: **2** ....

A: **3** ....

B: Yes, it's EX9624/TA12.

A: Is that Olivier Lesout?

B: **4** ....

A: And which catalogue have you got?

B: **5** ....

A: Right, what's the item number?

B: **6** ....

A: Oh, yes, the black corduroy trousers. What size would you like?

B: **7** ....

A: **8** ....

B: No, thanks. That's all.

A: **9** ....

B: By credit card. The number is 9866 0405 8341 7844, expiry 02/06.

A: **10** ....

B: Thanks. Bye.

a It's the winter catalogue.
b Yes, I have.
c How would you like to pay?
d It's P39/T143.
e Yes, I'd like to order some trousers, please.
f Can you give me your customer reference number?
g Thank you. Your order should be with you in three working days.
h Yes, that's right.
i They're €55, including post and packing. Would you like to order anything else?
j Medium, please.

Language study

Task 9 Talking about the future

Notice how we use the simple present when we talk about a fixed timetable:

The next train to Kyoto ***leaves*** *at 10.22.*

and we use the present continuous when we talk about future arrangements:

I'm meeting *her this evening.*

Now put the verb in brackets in the correct form in these sentences.

1 The Geneva flight .... at 5 p.m. (arrive)
2 I .... in the conference hotel next week. (stay)
3 The Paris train .... at 7 a.m. (leave)
4 We .... for all the delegates to receive a welcome pack on arrival. (arrange)
5 She .... the order details by email later this week. (confirm)
6 The sales conference .... at lunchtime tomorrow. (start)

Task 10 Nouns and verbs

Complete the table with the missing nouns and verbs. Use a dictionary to help you if necessary.

	Noun	Verb		Noun	Verb
1	....	deliver	9	....	reserve
2	....	inform	10	booking	....
3	cost	....	11	cancellation	....
4	....	enquire	12	quotation/quote	....
5	charge	....	13	....	arrange
6	....	confirm	14	translation	....
7	call	....	15	....	guarantee
8	suggestion	....	16	....	fly

Speaking

Task 11

Listen to the callers who ask you for information, phone and reference numbers. Pause the recording and answer them, using the information given.
Then listen to the correct way to say them.
You may listen to the recording first to help you.

Symbols used between numbers are spoken as follows:
/ = slash or stroke – = dash or hyphen

In website addresses:
/ = slash or forward slash . = dot @ = at
In British English: **0** = oh (or nought)
In American English: **0** = zero

1 0034
2 CH5067/39
3 dictionary.cambridge.org
4 A–1010 Wien
5 0082 2 7844076
6 bbc.co.uk/radio4
7 4381869E/06
8 floriane@pondnet.com
9 5797 4132 6581 2976
10 KL7954–326

You will find the tapescript on page 90.

Task 12

You would like to order this tent from a mail order catalogue. You have phoned the mail order company. Listen to what the customer adviser says. Pause the recording and respond.
You may listen to the recording first to help you.

Your details:
Name: Alex Harvey
Address: 74 Greenwood Road, Wimbledon, London SW19 1QU
Credit card number: 4954 6712 3695 3781, expiry 04/05

Lightweight ridge tent

- 2-person tent
- height – 1.5m

Item number: FD-4765
Price: £45

You will find the tapescript on page 90.

Task 13 Role play

Work with another student when you do this exercise. Agree which of you is Student A and which is Student B. Student A has information on this page, Student B on page 73. Sit back to back. Student A should now 'call' Student B. When you have done the calls once, change roles.

A1 You are a colleague of Martha Wong at the florist's shop Blooming Flowers. You are dealing with James Elliott's order: 15 red roses to be sent to his home address at 6 p.m. Call Mr Elliott and explain that there were very few red roses at the early morning flower market, and you have none left now in your shop. Ask him whether he would accept a different colour of rose for the bouquet: you have plenty of white and pink roses.

A2 You are a colleague of Serge Duval at BGX Computers. Serge has asked you to call Jennifer Sato at Kobayashi Components to explain that the Finance Department will not authorise payment for the order for hard drives by bank transfer. The Finance Director insists that the account should be paid in the normal way at the end of the month. You need to persuade Jennifer Sato to agree to this, and you need to confirm that the order will still be dispatched this week, as it is very urgent.

A3 You are a colleague of Alessandra Tauzia at Terracotta Italiana. Alessandra has asked you to call Brigitte Schmidt at the Edelweiss Garden Centre in Zurich, as there is a problem with the order. Explain that the stock of 35 cm pots, reference number AZ35, is low. You would be able to send 50 rather than the 120 pots they ordered. Ask if that is acceptable. You don't know when more stock is due in. Also, the Calabrian company have now told you that their big pots (75 cm, ref. no. CC75) are not guaranteed frost-proof. However, try to persuade Brigitte to order some, as you feel sure they would sell anyway.

5 Bookings and arrangements

Listening

Task 1

Listen to two phone conversations and complete the table.

Call	Name of travel agency	Destination	Alternatives
1			
2			

Task 2

Listen to the calls in Task 1 again. Decide if the statements about the calls are true (T) or false (F).

1 This is the first time Mike Wilkins has rung the travel agent asking for information. **T/F**
2 The travel agent has been to the places they are discussing. **T/F**
3 Hotel San Lorenzo is in the city centre. **T/F**

4 Helga Langendorf will be in Hong Kong for work. **T/F**
5 The price of the flights on the two airlines is very different. **T/F**
6 The travel agent will make the bookings immediately. **T/F**

You will find the tapescript on page 91.

What to say – what to expect

Read these useful sentences and make sure you understand them. Use a dictionary to help you if necessary.

Hotel reservations

Person calling

Could you tell me the price of a single room, please?

I'd like to book a double room for three nights, please.

Have you got any rooms available?

Are conference facilities available at the hotel?

How far is the hotel from the airport?

I'd like a quiet room with a balcony overlooking the sea.

I've got a room booked for tomorrow. I won't be arriving until about 11.30 p.m. You will keep the room for me, won't you?

I'm afraid I've got to change my booking. Something urgent has happened and I've had to change my plans.

Person called

A single room with shower or bath is €150, with breakfast included.

Would you prefer a shower or a bath?

All rooms are fully equipped with satellite TV, air-conditioning and Internet connection .

I'm afraid we're fully booked.

We haven't any double rooms left, but I can offer you a suite.

Could you send an email to confirm your booking?

Shall I send you the information pack about our conference facilities?

Task 3

Complete the sentences with words from the list below. Use each word once only.

1 Would you like a single or a .... room?
2 The .... time of flight JAL314 is 14.50.
3 Several .... fly the same route, so it's mainly a question of choosing the most .... time.
4 If you go on a .... flight, you're likely to pay more but it's often more convenient.
5 Don't forget: .... time is an hour before take-off.
6 They've decided to stay in an .... rather than a hotel.
7 Sorry to have .... you waiting.
8 The conference .... in the hotel are excellent.
9 All the .... have been made. You'll be .... from us soon.
10 I'd like to .... a suite with a balcony for two nights.

kept	scheduled	hearing	departure	facilities	convenient
book	apartment	double	arrangements	check-in	airlines

Task 4

Choose the best responses.

1 Do you want to book a scheduled flight?
 a Yes, the schedules are best.
 b Yes, it will be more convenient.
 c Yes, I like the airports.

2 What's the availability on the flight?
 a The flight is fully booked.
 b The flight is not available.
 c The flight will depart when the seats are available.

3 Can you send me an email as confirmation of the booking?
 a Yes, of course.
 b Please book by email.
 c I'll check the booking online.

4 Can I book three single rooms for our group tonight?
  a I'm sorry. You'll have to find another hotel.
  b I'm sorry. We don't handle package tours for groups.
  c I'm sorry. We're fully booked.

5 Is service included in the hotel rate?
  a Yes, you don't need to leave any tips.
  b Yes, your car will be serviced.
  c Yes, you will be served breakfast in your room.

6 Please book me on a flight at about 18.00.
  a Would you like a return ticket?
  b Will the 18.30 flight be OK?
  c What time does it arrive?

Task 5

Listen to the phone call and decide which message pad has the correct flight details.

1

2

3
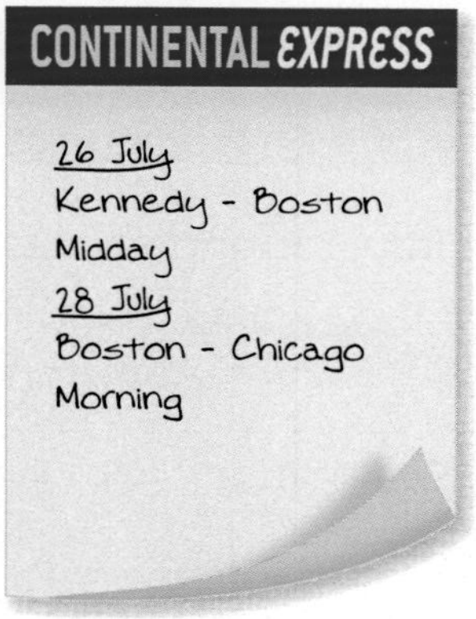

Listen again and complete the table.

Caller	
Hotel location	
Hotel name	
Booking dates	
Type of room	

You will find the tapescript on page 92.

What to say – what to expect

Read these useful sentences and make sure you understand them. Use a dictionary to help you if necessary.

Travel arrangements – air

Person calling

I'd like to book a seat on flight AZ514 from Paris to Frankfurt on 15 June.

How long is the flight from Berlin to Istanbul?

Is there a direct flight from San Francisco to South Korea?

Is there a connecting flight from Detroit to Miami?

How far is the airport from the city centre?

How long will the stopover in Bahrain be?

How much would it cost to take a taxi from the airport to the hotel?

Person called

I'm sorry there are no seats left on that flight.

Would you like a seat by the window or an aisle seat?

Would you like a business class or economy class seat?

There are five scheduled flights a day between Stockholm and Brussels.

Take a bus or the metro from the airport to the city centre.

I'm afraid there's no availability on the flight you wanted.

Travel arrangements – road and ferry

Person calling

What's the road like between Lisbon and Coimbra?

What's the best route to take over the Alps?

Could you please send me directions so I can find the office?

Is it easy to park in the city centre?

Do I need to book in advance to take the car on the ferry?

How long is the crossing from Bari to Dubrovnik?

Person called

The motorway is always very busy in the rush hour.

There will be major roadworks for the next few weeks, causing long delays.

I'll email you a map showing exactly where the hotel is.

The hotel has its own underground car park.

Due to rough weather, the ferry crossing scheduled to depart at 18.00 has been cancelled.

Travel arrangements – rail

Person calling

I'd like a return ticket from Geneva to Milan.

I'd like to take the overnight sleeper from Avignon to Paris.

I'm travelling on Eurostar, so I'll be in Brussels in plenty of time for the meeting.

Person called

Would you like to travel first class or standard class?

Would you like to be in a mobile-free carriage?

Would you like to book a table in the dining car?

Task 6

Listen to the phone conversation and complete the itinerary.

Details of visit

Name: Louis Gasquet
Company: Monteil SA , Lyon, France
Arrival date: **(1)** ....
Arrival time: **(2)** ....
Airline: Alitalia – AZ325
Transfer to hotel:
(3) ....
Hotel: **(4)** ....
No. of nights booked:
(5) ....

Meetings

First day: Lunch with key people at **(6)** .... ; meeting at 3 p.m. with the **(7)** ....; meeting at **(8)** .... with the Managing Director.
Second day: Whole day at **(9)** .... near Orvieto.
Back for **(10)** .... in Rome.
Third day: Return flight to Lyon.

Listen again and answer the questions.

11 Is this Louis Gasquet's first visit to the company in Rome?
12 What question does Louis ask Flavia at the end of the conversation?

You will find the tapescript on page 93.

Task 7

Listen to the phone conversation and complete the notes about changes.

Hotel Adlon Conference Centre

Booking details: Spectrum Technodesign

Monday 10 June

Rooms: 25 single, with shower/bath
Now: No change
Dinner in Linden Restaurant at 8 p.m.: 45 people, with 9 vegetarians
Now: **(1)** .... people, with **(2)** .... vegetarians

Tuesday 11 June

Rooms: 25 single, with shower/bath
Now: **(3)** .... single, with shower/bath; **(4)** .... double, with shower/bath
Dinner in Linden Restaurant at 8 p.m.: 15 people, no vegetarians
Now: **(5)** .... people, **(6)** .... vegetarians

Wednesday 12 June

Rooms: 25 single, with shower/bath
Now: **(7)** .... single, with shower/bath; **(8)** .... double, with shower/bath
Dinner in Linden Restaurant at 8 p.m.: 45 people, with 9 vegetarians
Now: **(9)** .... people, **(10)** .... vegetarians

You will find the tapescript on page 94.

Task 8

Complete the conversation with sentences from the list below. Use each sentence once only.

A: Iberia Airlines. Good morning. Can I help you?

B: **1** ....

A: **2** ....

C: Flight Reservations.

B: **3** ....

C: How can I help you, Ms Meier?

B: **4** ....

C: I see.

B: **5** ....

C: Are you flying business class?

B: **6** ....

C: Well, in that case, if there's a seat available on the plane, you'll have no problem. Do you know which flight you want?

B: **7** ....

C: **8** ....

B: Yes, please.

C: Right. Go to the Iberia desk at the airport at least 60 minutes before departure.

B: **9** ....

C: Yes, it's IB/0975/453.

B: **10** ....

C: Bye.

a But my conference is ending earlier and I'd like to take an earlier flight back.
b Hold the line, madam, and I'll put you through to Flight Reservations.
c Let's see ... yes, there are a few seats left. Shall I reserve one for you?
d Yes , IB3167 is the flight I'd like to take, at 15.45.
e That's fine, then. Thanks very much. Bye.
f Good morning. I'd like to change a flight booking, please.
g Hello. My name's Rosa Meier.
h And they'll change the ticket then? Is there a reference number?
i Well, I'm booked on an Iberia Airlines flight from Barcelona to Geneva this Friday at 18.40.
j Yes, I am.

Language study

Task 9 Probability and possibility

We often use *will*, *should* and *might* when we want to show how certain we are about what we are saying.

certain	The Managing Director *will* chair the meeting.
probable	Grace Lin *should* be back this afternoon.
possible	I *might* meet him later.

Now change the following sentences to show how certain you are. Example:

Your parcel is likely to arrive tomorrow.
*Your parcel **should** arrive tomorrow.*

1 I'm not sure if we'll visit Amsterdam on the way home.
2 It's likely that the consignment will reach you at the end of the week.
3 You'll probably get a good discount from the car company.
4 The discount is certain to be bigger if you book more than 50 seats.
5 The reference number is probably at the top of the page.
6 She's certain to call you before 12.00 tomorrow.

Task 10 Reporting questions

When you pass on a message, you will need to report different types of questions.

'Is the director satisfied with the arrangements?' (She asked)
*She asked **if/whether** the director was satisfied with the arrangements.*
'Why haven't you delivered the order?' (He wanted to know)
*He wanted to know **why** we **hadn't** delivered the order.*
'What will the discount be?' (She enquired)
*She enquired **what** the discount **would** be.*

Now report the following questions in a similar way.

1 'Have all the arrangements been made?' (They wanted to know)
2 'What is the reference number?' (He asked me)
3 'Is the hotel central?' (She enquired)
4 'How much does a double room cost per night?' (He wanted to know)
5 'How long will the conference last?' (She asked)
6 'Can I pay by credit card?' (He wondered)
7 'Have you booked the hotel?' (She wanted to know)
8 'Why have you changed the flight?' (He asked)
9 'What have they done with the files?' (She wondered)
10 'What time will the dinner start?' (He enquired)

Speaking

Task 11

Question tags are used at the end of sentences to ask for confirmation or agreement.

*She's French, **isn't she**?*

Question tags can be pronounced in two ways.
If they are spoken with a rising tone ↗, they are real questions: the speaker doesn't know the answer.
If they are spoken with a falling tone ↘, they are only asking for confirmation: the speaker knows the answer but wants to check it.

Listen to the examples that show the difference between the rising and falling tones.

Rising tone ↗	Falling tone ↘
*She's French, **isn't she**?*	*She's French, **isn't she**?*
*You've booked the room, **haven't you**?*	*You've booked the room, **haven't you**?*
*He doesn't like flying, **does he**?*	*He doesn't like flying, **does he**?*

Now add question tags to complete the sentences. Listen to the sentences to hear the difference between the rising and falling tones indicated by the arrows.

1 There aren't any seats left, ... ↗

2 She's already paid, ... ↘

3 We'll have to change the booking, ... ↘

4 The dinner was good, ... ↘

5 The flight's on time, ... ↗

6 You can ring them tomorrow, ... ↘

7 You liked that hotel, ... ↘

8 They haven't called us back, ... ↗

9 You'll make sure you're on time, ... ↘

10 You've got her mobile number, ... ↗

You will find the tapescript on page 95.

Task 12

You have received this email from Gregor Bachmann, postponing his visit to your company. He was due to come on Tuesday, 7 March and stay for two nights. You now need to ring the hotel and change the reservation you had made for Gregor. Listen to what the hotel receptionist. Pause the recording and respond. You may listen to the recording first to help you.

You will find the tapescript on page 96.

> I'm afraid I need to postpone my two-day visit to your company. The Managing Director of our parent company has decided to come for a series of meetings, and I obviously need to be here. Can we postpone my visit until the following week? I could come on Tuesday 14 and would stay for two nights, as agreed. Can you change the reservation at the Grand Hotel for me if the dates are OK with you?
>
> I look forward to hearing confirmation.
>
> Best wishes,
>
> Gregor Bachmann

Task 13 Role play

Work with another student when you do this exercise. Agree which of you is Student A and which is Student B. Student A has information on this page, Student B on page 74. Sit back to back. Student A should now 'call' Student B. When you have done the calls once, change roles.

A1 You work at the travel agency Choice Travel. Your colleague, Beth, has asked you to call Mike Wilkins about his reservation at the Hotel Reale in Barcelona. Unfortunately, the Hotel Reale is fully booked with a conference over the dates Mr Wilkins wants. You have contacted Hotel Del Norte and have made a provisional booking for him there instead – it is an excellent hotel, and is a similar price to the Hotel Reale. Try to persuade him to stay there – it is more central. You can recommend the Estrella restaurant for seafood, and the Meson Jordi restaurant for traditional cuisine.

A2 You are a colleague of Flavia Bellini. Flavia has asked you to call Louis Gasquet about a change of plan. The Managing Director is now unable to make the meeting arranged for Tuesday at 5 p.m. The meeting could be rearranged for Thursday at 5 p.m. You know that Louis wanted to go to an exhibition in Rome, if there was time. Find out whether he would be able to stay on in Rome until Friday, and what arrangements he would like you to make for him.

A3 You are a colleague of Jerry Kemp from Spectrum Technodesign. Jerry is very busy at the exhibition and has asked you to call Verena Fischer at the Hotel Adlon, where several staff from Spectrum are staying this week. Jerry has asked you to book a buffet lunch at the hotel tomorrow, and has given you the following notes.

> Buffet at 1 p.m. – Hotel Adlon
> Not sure how many people, about 35 – ask for advice on quantities; please check Spectrum won't have to pay for what isn't used.
> Ask for: fruit juice, mineral water, sandwiches and snacks (some vegetarian), cakes and tea/coffee.

6 A change of plan

Listening

Task 1

Listen to two phone conversations and complete the table.

Call	Caller	Person called	Original appointment	Reason for change	New arrangement
1					
2					

Task 2

Listen to the calls in Task 1 again. Decide if the statements about the calls are true (T) or false (F).

1 The visitor from Korea was originally going to come the previous week. **T/F**
2 Robert Manzini will be playing golf with clients on Friday. **T/F**
3 There won't be time for any lunch when they have the meeting. **T/F**
4 The Heron International office will reopen at 9 a.m. **T/F**
5 The caller will have to stay in bed for at least two days. **T/F**
6 The caller's colleague will take their list of proposals to the meeting. **T/F**

You will find the tapescript on page 96.

What to say – what to expect

Read these useful sentences and make sure you understand them. Use a dictionary to help you if necessary.

Making appointments

Person calling / Person called

How about meeting on Tuesday 21st at 11.00?
I'll just check my diary.
When would be convenient for you?
Could you manage one morning next week?
Shall we say Wednesday 29th at 3 o'clock?
I'm afraid I'm tied up all that day.
I'll pencil that in for now. Can you tell me soon whether we can go firm on that?
Sorry, I've already got an appointment then. Can we arrange another time?
Could you send me a fax / an email confirming all the details?
Would it be possible to postpone the meeting?
Sorry to be difficult, but something urgent has come up, and I'm not going to be able to make it on the day we'd fixed.
It looks as if everyone involved can manage Friday next week, so let's go for that.

Task 3

Complete the sentences with words or phrases from the list below. Use each word or phrase once only.

1 I can hardly .... you. It .... as if you're miles away.

2 My client had to make a .... change to her itinerary.

3 Let me just look at my .... Yes, I could certainly .... to come for a meeting on Tuesday afternoon.

4 Sorry, I can't .... it that day – I'm going to be .... with something else.

5 Several people can't come on Thursday – I think we'll have to .... the meeting.

6 Would it be .... for you if we come in two weeks' time?

7 There's a line through the diary that day – that means I have to keep it ....

8 I try to make sure I have at least half an hour between ....

9 Every .... seems to generate a lot of ....

10 Please could you email me .... of the arrangements?

appointments
convenient
confirmation
tied up
free
manage
paperwork
postpone
sounds
diary
hear
last-minute
make
meeting

Task 4

Choose the best responses.

1 Can we make an appointment?
a When are you free?
b Shall we make a reservation?
c Is it difficult for us to meet?

2 I'm going to have to postpone our meeting, I'm afraid.
a I don't want to postpone it.
b Don't worry, I'm sure we can find another date.
c What are you afraid of?

3 I'm tied up on Tuesday and Wednesday.
a Oh, I'm sorry to hear that.
b I'll come on Tuesday then.
c What about Thursday?

4 I'm glad we've finally managed to fix a date.
a So am I.
b Yes, we have.
c What about next Monday?

5 Is everything ready for the meeting?
a No, we're not there.
b It's all noted in my diary.
c Yes, I think so.

6 I hope this hasn't messed up your arrangements.
a Well, my desk is always in a mess.
b These things happen.
c Yes, I always arrange things like this.

Task 5

Listen to the phone conversation and complete the table.

Caller	Person called	Reason for calling	Appointment

You will find the tapescript on page 97.

What to say – what to expect

Read these useful sentences and make sure you understand them. Use a dictionary to help you if necessary.

Changing appointments

Person calling / Person called

Could we arrange another appointment?
How about the 7th rather than the 1st? Are you free then?
I'm sorry, I can't manage that day after all. Can we find another date?
I'm calling because I don't think I'll be able to come after all.
Let's fix another time then. Would it suit you if we postpone the meeting until next month?
There's been a change of plan and I'm afraid I'm going to have to rearrange things to try to fit everything in.
I've been double-booked, because my assistant was confused by the appointments written in my diary, so we'll have to change the time of our meeting.

Task 6

Listen to two phone conversations and complete the table.

	Caller	Dates booked	Room type	Special request
1 Hotel Saint-Jean				
2 Royal Western Hotel				

You will find the tapescript on page 98.

Task 7

Listen to the phone conversation in Part 1 and complete the notes on the message pad. Then listen to the phone conversation in Part 2 and complete the email.

Tarquin Services

Annual Dinner
Date **(1)** ....
Barbara Zimmermann called – problem:
(2) ....
Alternative dates:
(3) ....
Hotel booked :
(4) ....
Time: **(5)** ....

To: All
Subject: Tarquin Services Annual Dinner

Dear Colleagues,
As you know, we had previously arranged to hold our Annual Dinner on **(6)** .... at the **(7)** .... Hotel. Due to unforeseen circumstances, we have had to postpone it by a few days. It will now be held at the Hotel **(8)** .... on **(9)** .... at **(10)** .... p.m. A room has been booked for you in the hotel that night.

I hope this will not inconvenience you. Please send confirmation that you will be able to come. I look forward to seeing you all there.
Best wishes, Barbara Zimmermann
Marketing Manager

Listen again to Part 1 and Part 2 and answer the questions.

11 Why can't the speaker come on the date they'd arranged?
12 Why does Henri ask Barbara if she will email everyone?

You will find the tapescript on page 99.

Task 8

Complete the conversation with sentences from the list below. Use each sentence once only.

A: **1** ....

B: Oh, fine really. And you, Sara, how are you?

A: **2** ....

B: Well, we're sending a film crew to do a short piece in Malaysia.

A: **3** ....

B: You guessed!

A: **4** ....

B: For two weeks, leaving at the end of next month.

A: **5** ....

B: That's right.

A: **6** ....

B: Good. I'll start getting things organised then.

A: **7** ....

B: What's the problem? We'll pay you your normal rate.

A: **8** ....

B: And then you'll be able to confirm?

A: **9** ....

B: Yes, I'll do that, Sara. I hope you can get everything arranged.

A: **10** ....

a Fine, thanks. What can I do for you?
b Hold on a minute. It's not quite definite yet.
c I'm sure I can. Thanks for calling, Alex.
d No, it's not that. I'll have to make some arrangements first.
e Hello, Alex, good to hear you. How are things?
f So only a fortnight?
g When would the trip be?
h Well, I'll just check … yes, that should probably be OK.
i Yes, if I haven't confirmed by the end of the week, ring me again.
j Ah, let me guess. You need a sound recordist.

Language study

Task 9 Future possibilities

Study these examples of how to talk about future possibilities.

What would happen if you couldn't keep the appointment? (I/telephone/apologise)
***I would telephone** and apologise **if I couldn't keep** the appointment.*
What would happen if Sun Jim Kim wasn't offered the job? (he/apply elsewhere)
*If Sun Jim Kim **wasn't offered** the job, **he would apply** elsewhere.*

Now complete the answers to these questions in a similar way.

1 What would you do if the hotel was fully booked? (I/find another hotel)
If ....
2 What would happen if the speaker couldn't come to the conference? (we/look for a replacement)
If ....
3 What would you do if your client couldn't come to the meeting? (we/postpone the meeting)
We ....
4 What would be the effect if the computers crashed? (it/be/a disaster)
It ....
5 What would Markus do if the sales figures were well below the target? (Markus/resign)
If ....
6 What would you do if you missed your flight? (I/take/the train)
I ....
7 What would Anne Marie do if she forgot to go to her dental appointment? (she/ring/apologise)
If ....
8 What would happen if you were ill that week? (Catherine/stand in for me)
If ....

Task 10 Phrasal verbs

We often use phrasal verbs in conversation. Choose the correct adverb or preposition in brackets to complete these phrasal verbs.

1 Sorry, I can't make the meeting this afternoon. Could we put it ________ until tomorrow? (up/forward/back)

2 I'll get in touch ________ Matt Sefton and see if we can change the date. (with/by/for)

3 I'd like to go ________ the report before the meeting. (in/through/at)

4 My diary is pretty full, but I could fit ________ a lunchtime appointment on Thursday. (on/in/at)

5 It's too late to change everything now – let's stick ________ our original plan. (with/on/up)

6 Send Louis an email with the details and copy me ________ . (on/by/in)

7 Thanks a lot for sorting it ________ . (through/over/out)

8 The only suitable dates are the 4th and the 12th. Let's go ________ the 4th. (at/for/by)

Now match each of the phrasal verbs to one of the definitions from the list below.

a choose
b read carefully
c find time in a busy period
d send a copy to someone
e deal with a problem
f contact someone
g postpone
h stay with / not change

Speaking

Task 11

Listen to the callers. Pause the recording and answer their questions, using the information given.
You may listen to the recording first to help you.

1 9–11 July
2 PB/9534–06
3 Heydenfeldt
4 0039 011 864 4360
5 wilson@transdeal.netvigator.com
6 07998 652714
7 17 April
8 Polozova

You will find the tapescript on page 100.

Task 12

Here is a page from your diary for 2 February. Philippe Lamoine phones to change the meeting with you at 10 a.m. because he has a dentist's appointment. Have a conversation with him and arrange a time for another meeting later in the day. Listen to what he says. Pause the recording and respond.
You may listen to the recording first to help you.

2 February	Tuesday
9.00	*Phone Taiwan office*
10.00	*Philippe Lamoine - here*
11.00	*Carmen Lee / Publicity*
12.00	*Lunch with Hana Dankova*
1.00	
2.00	*Warehouse*
3.00	
4.00	
5.00	*Sales meeting - room 308*

You will find the tapescript on page 101.

Task 13 Role play

Work with another student when you do this exercise. Agree which of you is Student A and which is Student B. Student A has information on this page, Student B on page 75. Sit back to back. Student A should now 'call' Student B. When you have done the calls once, change roles.

A1 You are Professor Stephanie Odermatt, a specialist in company management. You regularly give speeches at conferences and dinners. Your normal fee is €2,000 plus travel expenses. You already have several bookings in November, and you will be away on holiday for the last week, but you would be prepared to fit in one more speech.
Here is a summary of your availability in November:
Monday 3, Friday 7, Wednesday 12, Tuesday 18, Thursday 20.

A2 You are a colleague of David Harper, who has asked you to arrange a meeting with Amanda Walters. Call her to ask her what day next week is suitable for a meeting. Here is David's diary for next week.

Monday

3 pm Frank Kaufmann here

Tuesday

9-12 meeting with MD

Wednesday

12-2 lunch with area manager

Thursday

10 am team meeting
2 pm video conference

Friday

2 pm finish report for Monday's meeting

7 What's the problem?

Listening

Task 1

Listen to two phone conversations and take notes on the message pads.

1

Caller

Address

Notes

Penta Magazines

2

City Pizzas

Caller ....

Order no: 10964/32 (Monday 17 July)

.... pizzas

Delivery: 21 July, 12.30, Downtown Studio

Notes

Task 2

Listen to the calls in Task 1 again. Decide if the statements about the calls are true (T) or false (F).

1 The customer adviser checks the caller's postcode. **T/F**
2 The magazine comes out every month. **T/F**
3 The magazine probably got lost in the post. **T/F**

4 The caller hoped the missing pizzas were about to arrive. **T/F**
5 The caller couldn't understand how the mistake had been made. **T/F**
6 City Pizzas won't charge the caller for the total number of pizzas delivered. **T/F**

You will find the tapescript on page 101.

What to say – what to expect

Read these useful sentences and make sure you understand them. Use a dictionary to help you if necessary.

Checking up on problems

Person calling	Person called
I still haven't received the order.	What seems to be the trouble?
I can't understand why there's been a delay.	Let me check the records.
I've been waiting for seven weeks now.	It's obviously our mistake.
Can you check up on it, please?	I'm very sorry about that.

Task 3

Complete the sentences with words from the list below. Use each word once only.

1 I'm phoning you about a .... matter.

2 You can't have packed it properly; the package was already .... when it was .... to us.

3 We expected a much higher standard of ....

4 I'm ringing to say how .... I was by the final result.

5 I can only .... on behalf of the company.

6 I'm afraid there's been a .... ; your order was dispatched to the wrong address.

7 We haven't received the parts from our supplier, so there's inevitably a .... in .... your order.

8 We're very sorry for the ....

processing	mix-up	inconvenience	delay	service
apologise	delivered	disappointed	serious	damaged

Task 4

Choose the best responses.

1 Can you explain why the consignment has got stuck at customs?
  a So we can collect it, can we?
  b We're not sure yet why it has been delayed.
  c You mean they've stamped it.

2 It will be delivered by the courier company we always use.
  a It's very fragile.
  b Will they be here soon?
  c Are they reliable?

3 I'll have to make a complaint.
  a Yes, please do.
  b When can you make it?
  c If you see them, tell them.

4 I'm sorry to have to report that it isn't acceptable.
  a Who's done it?
  b We need better service.
  c You'll have to put in a complaint.

5 There's been a bit of a mix-up.
 a Why did you disturb it?
 b I'm sorry everything is in the wrong place.
 c What's the problem?

6 I'll try to get things moving as quickly as I can.
 a The sooner you go, the better.
 b I'd appreciate that.
 c Don't move things without telling me.

Task 5

Listen to the phone conversation in Part 1 and complete the table. Then listen to the phone conversation in Part 2 and complete the notes on the message pad.

Caller	Reason for complaint	Next step

Quicklink Couriers
Fast Fax Central Service Department - Complaint (9 March)
Ref. No. RZ2984/W56
We collected fax machine from Fast Fax Central on: **(1)** ....
Driver tried to deliver fax machine on: **(2)** ....
Did driver leave card for customer? **(3)** ....
Now we must ring customer to arrange convenient time for
(4) ....

You will find the tapescript on page 103.

What to say – what to expect

Read these useful sentences and make sure you understand them. Use a dictionary to help you if necessary.

Making and handling complaints (1)

Person calling
I'm afraid I have to make a complaint.
It's very inconvenient.
The standard of service was unacceptable.
We should have been warned there was a problem.
I think we'll have to ask for a refund.
What are you going to do about it?

Person called
I'm very sorry to hear that.
I'm very sorry about the delay.
I'll find out what has happened and ring you back.
I shall make a full investigation into what went wrong.
I appreciate your position.
I can only apologise.

Task 6

Listen to the phone conversation in Part 1 and decide which email is the best record of what was said. Then listen to the phone conversation in Part 2 and answer the questions.

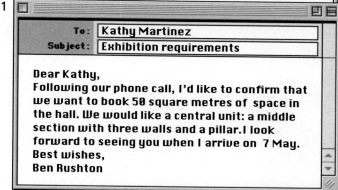

1

To: Kathy Martinez
Subject: Exhibition requirements

Dear Kathy,
Following our phone call, I'd like to confirm that we want to book 50 square metres of space in the hall. We would like a central unit: a middle section with three walls and a pillar. I look forward to seeing you when I arrive on 7 May.
Best wishes,
Ben Rushton

2

To: Kathy Martinez
Subject: Exhibition requirements

Dear Kathy,
Following our phone call, I'd like to confirm that we want to book 40 square metres of space in the hall. We would like a central unit: a middle section with three walls but no pillar. I look forward to seeing you when I arrive on 7 May.
Best wishes,
Ben Rushton

1 Where is Ben Rushton calling Kathy Martinez from?
2 What is wrong with the space Ben has been given?
3 Whose fault is it that the mistake was made?
4 How does Kathy resolve the problem?
5 By way of an apology, Kathy offers to help Ben in two ways. What are they?

You will find the tapescript on page 104.

Task 7

Listen to the phone conversation and complete the table.

Caller	Place called	Complaint 1	Complaint 2	Complaint 3

Listen again and answer the questions.

1 How long was the conference held by City Management Services at the hotel?
2 How many people from City Management Services stayed at the hotel?
3 Which of the three things the caller complains about was the most important and why?
4 Did the conference manager know about the problems City Management Services had experienced before the caller rang to complain?
5 By how much will the invoice be reduced by way of an apology?

You will find the tapescript on page 106.

Task 8

Complete the conversation with sentences from the list below. Use each sentence once only.

A: **1** ....

B: Hello. Could you put me through to Oliver Chan, please?

A: **2** ....

B: Oh dear. Do you know when he'll be back?

A: **3** ....

B: Well, I'm afraid we're having problems with some of the parts you've sent us.

A: **4** ....

B: Yes, that sounds like the right department.

A: **5** ....

C: Technical Services. Jorge Casso speaking.

B: **6** ....

C: Good morning, Ms Peuser. How can I help you?

B: **7** ....

C: Do you have the code numbers for them, Ms Peuser?

B: **8** ....

C: Yes, that's a specially designed range, I think.

B: **9** ....

C: What seems to be the problem with them?

B: **10** ....

a I'll put you through to Jorge Casso then.
b I'm afraid he's away from the office.
c Yes, they're all from the DE1065 range.
d Oh, hello, my name's Claudette Peuser, I'm from Tyson-Scotts.
e Well, I'm afraid they're not exactly the right dimensions.
f CTK Electronics. Good morning.
g Not until next week, I'm afraid. Can somebody else help you?
h That's right.
i Well, we bought some special parts from you and …
j I see. Technical Services should be able to help you.

Language study

Task 9 Apologising

It is sometimes necessary to apologise because someone has not done something that they *should have done*. Study this example.

The consignment was delayed at the customs. (send/more documentation)
*I'm sorry. We **should have sent** more documentation.*

Now apologise in a similar way in these situations.

1 The hotel rooms weren't clean when the guests arrived. (check/ready)
2 Nobody told us there was a problem. (we/warn you/about delay
3 You sent our parcel to the wrong address. (check/your order)
4 Two of the three boxes arrived damaged. (pack/properly)
5 There was no instruction manual. (put/in the box)
6 The contract wasn't included with everything else. (check/envelope/before sending it out)

Task 10 Getting things done

You will often need to say that you will get another person to perform a service for the person you are talking to. Study this example.

The room you put me in isn't clean.
*I'll **have it cleaned** for you.*

Now change these sentences in a similar way.

1 Are you sure this invoice is correct? (check)
2 The photocopier isn't working properly. (fix)
3 There may be some letters for me. (forward)
4 I'd like some information about the latest model. (send)
5 We need the parts as soon as possible. (dispatch at once)
6 I've left my luggage in the conference room on the 10th floor. (bring down)

Speaking

Task 11

Study the table. Then say the figures and calculations aloud. Listen to check you have said them correctly. You may listen to the recording first to help you.

=	equals, is equal to, makes	1,204	one thousand two hundred and four
+	plus, and, add	3.75	three point seven five
–	minus, less, take away	$^1/_2$	half
×	multiplied by, times	$^3/_4$	three quarters
÷	divided by	$^5/_8$	five eighths

1 $^1/_2 + ^3/_4 = 1^1/_4$
2 $4.2 \times 3 = 12.6$
3 $36 \div 9 = 4$
4 17,506
5 78.5%
6 $391 - 62 + 148 = 477$
7 $^{11}/_{16} - ^5/_8 = ^1/_{16}$
8 $7.3 + 29.2 = 36.5$
9 $43 \times 5 = 215$
10 $2,640 \div 8 = 330$

You will find the tapescript on page 107.

Task 12

You work for Zanda Electrics. Here is a copy of an order you sent to Expedia Electronics and the invoice you have just received. Phone Angela Rusita of Customer Services to complain. Listen to what she says. Pause the recording and respond.
You may listen to the recording first to help you.

Zanda Electrics

Order
To: Expedia Electronics
Date: 16 March
Order no. 8451
Please supply:
15 x model EE721 clips at £4.75 each

Expedia Electronics

Invoice No.	54391
	27 March
To:	Zanda Electrics
Re:	Order No. 8451
	50 x model EE721 clips at €4.75 each
Total due:	€237.50

You will find the tapescript on page 107.

Task 13 Role play

Work with another student when you do this exercise. Agree which of you is Student A and which is Student B. Student A has information on this page, Student B on page 76. Sit back to back. Student A should now 'call' Student B. When you have done the calls once, change roles.

A1 You are a colleague of Tony Martin at Fast Fax Central Service Department. Tony has had to go home suddenly and has asked you to make an urgent call on his behalf. He had promised to ring Vera Steiner back as soon as possible to explain why there had been such a delay in receiving her repaired fax machine. Tony had discovered that it was Quicklink Couriers' fault. You know Ms Steiner is very angry and has been very inconvenienced. Explain that Quicklink Couriers will call her to apologise and fix a convenient time for delivery.

A2 You work at Quicklink Couriers, and are in charge of customer liaison. You have learnt about the problem when one of the new drivers forgot to leave a card saying he'd tried to deliver a repaired fax machine to a customer of Fast Fax Central over a month ago. You now have to ring the customer, Vera Steiner, to apologise for the driver's mistake, and to arrange a convenient time for the fax machine to be delivered.

A3 You are a colleague of Ben Rushton, who is busy with a customer at the moment. It's the second day of the exhibition, and Ben has asked you to ring Kathy Martinez, of Exhibition Organisers, who have made all the arrangements. Ben had ordered a buffet lunch for 35 people at the stand, to be ready at 12.30. It is now 1 p.m. and there is still no sign of any food or drink, or any waitresses. You are getting increasingly concerned, as you have invited several important clients to the buffet, and they have already arrived.

8 Handling complaints

Listening

Task 1

Listen to two phone conversations and complete the table.

Call	Company/Person calling	Company/Person called	Reason for calling	Action
1				
2				

Task 2

Listen to the calls in Task 1 again. Decide if the statements about the calls are true (T) or false (F).

1 This is not the first time the two people have spoken. **T/F**
2 The company calling has sent three emails about the problem before making this phone call. **T/F**
3 The person called says they have had problems with the computers in the office. **T/F**

4 When the caller first rang the airline office, she found a message on the answerphone. **T/F**
5 The caller's flight landed on time. **T/F**
6 The airline will phone the caller to arrange for the luggage to be delivered. **T/F**

You will find the tapescript on page 108.

What to say – what to expect

Read these useful sentences and make sure you understand them. Use a dictionary to help you if necessary.

Making and handling complaints (2)

Person calling
I'm calling to make/register a complaint.
I'm not satisfied with the service.
This simply isn't good enough.
Can you tell me what's going on?
Can I count on that?
I sincerely hope I don't have to ring you again about this.

Person called
It looks as if there's been an error at this end.
I'll have to look into this.
It's obviously a major slip-up.
I'm afraid the repair centre is running behind schedule.
One of our suppliers has let us down.
I'm very sorry for the inconvenience.
We'll make it top priority.

Task 3

Complete the sentences with words from the list below. Use each word once only.

1 I'm phoning because your payment is ....

2 If you let us have all the .... , we can .... out the problem.

3 It's such a .... not having my luggage here.

4 Somebody in the department has made a big .... and hasn't completed the job.

5 Now we've got your .... check number, we should be able to .... your missing luggage quite quickly.

6 The person who normally .... with these matters is off .... at the moment.

7 I'm very .... I can only .... on behalf of the company.

8 We need to have .... that our instructions have been followed.

apologise	trace	mistake	sick	nuisance	confirmation
sorry	details	baggage	overdue	sort	deals

Task 4

Choose the best responses.

1 I think there's been a slip-up somewhere.
a What message?
b Where's the note now?
c What sort of mistake?

2 There may have been a mistake at our end.
a So you've found it at last.
b I don't know when the mistake was made.
c Well, it certainly wasn't our fault.

3 I hope you can sort it out.
a It's difficult to arrange.
b I'm sure we'll find out what went wrong.
c What sort do you want?

4 My luggage is missing. It's a real nuisance.
a Yes, I'm sure it's most inconvenient.
b So you have a lot of cases?
c Yes, I'm sorry we don't have any here.

5 If you give me the check number, we'll trace your luggage for you.
a It's already labelled.
b I don't need a duplicate number.
c It's BA 0561354.

6 I'm sure your complaint is justified.
a Yes, I've just managed it.
b Yes, it's not the first time either.
c Yes, I've just made it.

Task 5

Listen to the phone conversation and complete the notes.

Complaint Record

Customer: **(1)** ....

Customer reference: **(2)** ....

Notes: faulty **(3)** ....

(flickering + purple stripes) – collected for repair on **(4)** ....

– received at **(5)** .... 16 May – she expected it back within a week – urgent repair needed now – must be delivered to her by the end of **(6)** ....

You will find the tapescript on page 109.

What to say – what to expect

Read these useful sentences and make sure you understand them. Use a dictionary to help you if necessary.

Making and handling complaints (3)

Person calling

We're very unhappy with the arrangements.
This really isn't acceptable.
That may be true, but it doesn't help us right now.
I want you to get this sorted out now.
I can't afford to be let down again.

Person called

We're normally very reliable.
We've never had this sort of problem before.
This certainly shouldn't have happened.
I'll look into the matter immediately.
I'll deal with it personally.
I can't apologise enough.
Thank you for telling me about it.

Task 6

Listen to the phone conversation and complete the table.

Caller/Company	Company called	Reason for call	Action

Listen to the phone conversation again and answer the questions.

1 Which day was the booking made?
2 Which day was the mini-cab booked for?
3 Where was the caller going?
4 How did the caller get to the meeting?
5 Will the caller use the mini-cab company in the future?

You will find the tapescript on page 111.

Task 7

Listen to the phone conversation and complete the table.

Caller	Company called	Reason for call	Problems

Listen again and decide if the statements are true (T) or false (F).

1 The caller hasn't come to the UK direct from the USA. **T/F**
2 The caller made the booking online. **T/F**
3 The owner of the apartment apologised when he arrived. **T/F**
4 The caller wants to stay in the apartment tonight. **T/F**
5 The caller will have to pay for the taxi herself. **T/F**

You will find the tapescript on page 112.

Task 8

Complete the conversation with sentences from the list below. Use each sentence once only.

A: Zorivos Pharmaceuticals.

B: **1** ....

A: A little. How can I help you?

B: **2** ....

A: Yes, that's right. Who would you like to speak to?

B: **3** ....

A: Please hold the line while I connect you.

C: **4** ....

B: Good morning. This is Javier Perez. I'm calling from Solero Farma in Spain.

C: **5** ....

B: I'm calling about the work we're doing on your new anti-histamine preparation.

C: **6** ....

B: And I made a provisional arrangement to come to your offices this Friday for a meeting with Ms Lindenberger.

C: **7** ....

B: And I'm calling to confirm the arrangements now.

C: **8** ....

B: Why, what's the problem?

C: **9** ....

B: Oh, that's a bit annoying, as it was the only day I could manage this week, and I'm away next week.

C: **10** ....

a I'm very sorry she didn't warn you – it was a last-minute decision when a crisis came up.
b Oh, I'm sorry, that won't be possible now.
c Oh, yes, Eveline Lindenberger said she was using a consultancy in Spain.
d Do you speak English?
e That is Zorivos Pharmaceuticals, isn't it?
f I see.
g Mrs Lindenberger, please.
h Good morning, Mr Perez.
i Eveline Lindenberger's phone. Aline Rosch speaking.
j I'm afraid Eveline had to go to Chicago for an urgent meeting yesterday.

Language study

Task 9 Fault diagnosis

When we are talking about faults, we often use modal auxiliary verbs:
may/might/could **– to list possible reasons (affirmative)**
may/might **– to list possible reasons (negative)**
should/ought to **– to talk about what we expect to happen**
can't **– to exclude various reasons**
must **– when we decide what the reason for the fault is**

Example:

I can't get a dialling tone on my phone. It *may* be because there's a fault in the phone, or it *could* be in the phone line, or I *might* not have plugged the phone in. So I check the plug, but it's in the socket, so it *can't* be that. The phone company says the line is OK, so the phone *ought to* work, but it doesn't. The fault *must* be somewhere in the phone itself.

Complete the fault diagnosis in a similar way. (In some cases there is more than one possible answer.)

My car won't start one morning. Why not? There are a number of possible explanations. It (**1**) .... be the battery. It (**2**) .... be the plugs. I check them both, but they're OK, so the car (**3**) .... start, but it doesn't. I had a new starter motor put in last month, so it (**4**) .... be that. Then I notice that the needle in the fuel gauge is pointing at empty. It (**5**) .... be the fuel. The fuel tank (**6**) .... be empty. If I put some fuel in it, the car (**7**) .... start. The fuel gauge (**8**) .... be faulty, but I don't think it is.

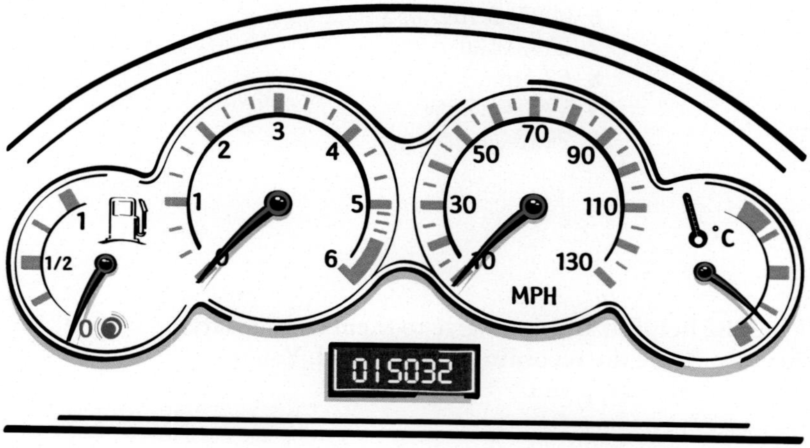

Task 10 Nouns and verbs

Complete the table with the missing nouns and verbs. Use a dictionary to help you if necessary.

	Noun	Verb		Noun	Verb
1	....	announce	9	description	....
2	schedule	....	10	preparation	....
3	cooperation	....	11	....	delay
4	....	apologise	12	arrival	....
5	....	prefer	13	....	recommend
6	statement	....	14	pleasure	....
7	....	complain	15	transmission	....
8	....	refer	16	....	depart

Speaking

Task 11

Listen to the callers. Pause the recording and answer their questions, using the information given. You may listen to the recording first to help you.

1 asap
2 Trifonidou
3 23.5 tonnes
4 15 July
5 D–30161
6 00971 3 7619 455
7 ETA 17.20
8 €2.5m
9 VAT = 17.5%
10 Paulhaguet

You will find the tapescript on page 113.

Task 12

Two people call you. Be helpful, and apologise to them if necessary. Listen to what they say. Pause the recording and respond. You may listen to the recording first to help you.

You will find the tapescript on page 113.

Task 13 Role play

Work with another student when you do this exercise. Agree which of you is Student A and which is Student B. Student A has information on this page, Student B on page 76. Sit back to back. Student A should now 'call' Student B. When you have done the calls once, change roles.

A1 You are a colleague of Rob Godwin at Flyfast Airlines. Rob has asked you to ring Marina Donato. You have managed to track down her missing luggage, but unfortunately you will not be able to get it to her hotel until tomorrow afternoon. Call her to apologise for the delay, and explain that Flyfast Airlines will offer her some compensation for the expenses she will incur because her luggage is missing.

A2 You work in the service department of Bell-Watson Computers. Call Bettina Seitz to tell her that her computer monitor has now been repaired. Try to arrange a time for it to be delivered.

A3 You work at Superior Accommodation. You call Mr Wainwright, the 'unreliable' owner of the apartment that Ms Clayton had booked through your agency. You have several points to complain about:

- He didn't have his mobile switched on, so the clients and the agency couldn't contact him.
- He kept the clients waiting outside the apartment.
- The apartment hadn't been cleaned in readiness for the clients.
- Superior Accommodation is having to pay for the clients to stay overnight in a hotel while the apartment is made ready for them.
- You are only willing to deal with reliable owners, so you will no longer advertise Mr Wainwright's apartment on your website.

Student 'B' Role plays (Task 13)

1 How can I help you?

B1 You are a colleague of Richard Dawson. Richard is away on holiday for a week. Make a note of the caller's message, and say Richard will deal with it on his return.

B2 You are a colleague of Mark Wheeler. Explain to the caller that Mark Wheeler is in a meeting. Ask if you can help the caller. Explain that you have the price list: the price of plugs (reference number MS/74/07) is £24.99 for ten.

B3 You are a colleague of Richard Dawson. Richard was on holiday last week, and is now in a meeting, and can't be disturbed. Explain that he is very busy. Ask the caller how you can help, and explain that you will ring back when Richard has had time to give you the information.

2 Hold the line, please

B1 You don't know the person the caller wants to speak to. Has he/she got the right number? Your number is 3486 5902.

B2 You work in the same office as Yoshida Tokuko, who is away for three days. Find out what the caller wants. You think the date is OK, but tell the caller that you will ask Yoshida Tokuko to confirm the arrangement on his return.

B3 You are a colleague of Teresa Lombardo. Teresa is away at a conference. Find out what the caller wants. Explain that you don't know anything about it, but you will do your best to find out. You promise to call the caller back with some information later today.

3 Making enquiries

B1 You are a colleague of Daniel Evans at Capital Investment Services, but you are not an expert on advising customers about investing their money. You don't want to lose the business of the customer who calls, but you are only able to give him/her limited information. Take down the customer's details and explain that you will pass on all the information to Daniel and ask him to ring the customer tomorrow.

B2 You are a colleague of Annabel Davies at Globe Travel Agency. There are flights from London to Lima every day with a stopover in New York. It is possible to stop over in Houston, Texas and then in Mexico City en route to Lima, but as this involves two airlines, the price would be £100 more. There are only two direct flights a week from Lima to Port of Spain, but there are daily flights from Trinidad to London. Take notes on what the customer wants, and explain that you will have to look up all the prices, and will call back.

B3 You are a colleague of Eva Frei, the manager at the International Shop in Berlin. You know the conference centre have confirmed that they will place an important order for tableware made by Bebbington Porcelain. They have chosen the Blue Room Collection on the understanding that everything will be delivered within six weeks. You know they were close to placing an order with a rival company, but the special discount Eva Frei was able to negotiate persuaded them to choose the tableware made by Bebbington Porcelain. Eva Frei is away and you are in charge. You don't want to lose the order.

4 Placing an order

B1 You are a colleague of James Elliott, who is in a meeting all day. He has asked you to deal with all his phone calls. You know it's James's partner's birthday and that James has ordered some flowers for her.

B2 You are a colleague of Jennifer Sato at Kobayashi Components. Jennifer is away on holiday all this week, and you are in charge. You know that Serge Duval at BGX Computers placed an order for 1,500 CM 25 hard drives, and that Jennifer agreed to reduce the unit price to $88 rather than $89 on condition the order was paid for immediately by bank transfer. You can't agree to any changes to this until Jennifer is back next week.

B3 You are a colleague of Brigitte Schmidt at the Edelweiss Garden Centre in Zurich. You know Brigitte has placed an important order with Terracotta Italiana. Note down the information you are given, and explain that Brigitte will be in touch with Terracotta Italiana when she is back in the office tomorrow. You would like to place an additional order: 100 30 cm pots, ref. no. AZ30.

5 Bookings and arrangements

B1 You are a colleague of Mike Wilkins, who is away from the office. You know he was looking forward to staying at the Hotel Reale on his forthcoming trip to Barcelona. He has planned several important meetings with business contacts, so the hotel and its location are both important. He has asked you to find out about a good restaurant in Barcelona, and to book a table for five on 19 May. He has asked you to make all decisions for him in his absence.

B2 You are a colleague of Louis Gasquet, who is away this week. You have his office diary. You know he is due to go to Rome on Tuesday, 25 October and that several meetings have already been arranged by Flavia Bellini. Make notes on the diary, so you can tell Louis about any changes to the Rome trip when he is back in the office.

Tuesday 25 October

0815 AZ325 flight Lyon - Rome, arrival 09.30
3 p.m. - meeting with Sales Director
5 p.m. - meeting with Managing Director

Wednesday 26 October

All-day visit to main factory near Orvieto

Thursday 27 October

08.00 AZ305 flight Rome - Lyon (change this?)
Any free time to go to the exhibition in Villa Borghese?

Friday 28 October

In office all morning
Leave office at 4 p.m. - going to Nice for the weekend

B3 You are a colleague of Verena Fischer at the Hotel Adlon. She has asked you to deal with Jerry Kemp from Spectrum Technodesign, who have booked several rooms and dinners in the hotel this week. You are used to dealing with orders for buffet meals, finding out the exact requirements, and giving advice. The hotel policy is to charge for all food that is ordered but not consumed, but not to charge for bottles that are not opened.

6 A change of plan

B1 You are a friend of Henri Julien. He has recommended Professor Stephanie Odermatt as a speaker at the conference you are helping to organise. The conference will last from Wednesday 5th to Sunday 9th November, and you would like her to speak to the delegates one day. The maximum fee that can be paid for a speaker is €1,750 plus travel expenses. Call Professor Odermatt to arrange for her to speak at the conference.

B2 You are a colleague of Amanda Walters. Amanda has told you that David Harper will probably call to arrange a meeting for next week. She's asked you to arrange it. Here is Amanda's diary for next week.

Monday

Holiday

Tuesday

2 p.m. onwards – Interviewing candidates for sales job

Wednesday

a.m. Must keep free to prepare talk
2 p.m. Dentist appointment
3 p.m. Ring Mexico office

Thursday

12.30 Lunch with Sam
p.m. Working at home

Friday

1 p.m. Meeting with Sales Team

7 What's the problem?

B1 You are staying in Vera Steiner's apartment all this week. Vera has had to go out for a short time to send some faxes and post some parcels. You overheard the phone call when Vera rang Fast Fax Central to find out why her fax machine was taking so long to be repaired. She has already explained to you how frustrated she is at not having a fax machine at the moment.

B2 You are the same person as in B1, and Vera is still out. Make sure the caller knows how inconvenient it has been for Vera not having her fax machine for so long. Explain to the caller that Vera is self-employed and that she is dependent on technology in order to earn her living. You know that Vera will be away for the next two days, and you have made arrangements to do several things this week yourself. You will need to look in your own diary to arrange a time for delivery.

B3 You are a colleague of Kathy Martinez, at Exhibition Organisers. It is the second day of the exhibition, and you are all working flat out. You don't know where Kathy is at the moment (1 p.m.). You know the catering staff have been under a lot of pressure, and that some of the exhibitors have not received exactly what they ordered. Your job is to try to sort out all the problems and apologise for anything that has gone wrong.

8 Handling complaints

B1 You are Marina Donato, and you are furious that Flyfast Airlines lost your luggage after your flight from Genoa. When they ring you, make sure they understand how inconvenient it is for you not having your luggage. You have had to buy some toiletries, and have borrowed clothes from a colleague. You don't intend to use Flyfast Airlines again.

B2 You are a friend of Bettina Seitz, who is out at the moment. You know all about the problems Bettina has had getting her computer monitor repaired, and the inconvenience it has caused her. She has had to rent a monitor in order to work, and you think she should get some compensation for the expenses she has incurred. Arrange a time for the monitor to be delivered.

B3 You are Mr Wainwright. Superior Accommodation advertise an apartment you own and rent it out to visitors. The income you receive from them is very important to you. You have had a lot of problems recently.

- You had forgotten which day the American visitors were due to arrive.
- The cleaner didn't clean the apartment because she was ill. She didn't ring you to explain the problem until this morning.
- You didn't know what a dreadful mess the previous visitors had left the apartment in.

  You are very apologetic.

Tapescripts

1 How can I help you?

Task 1 and 2

1

Anne Whitworth:	Hello.
David Bartlett:	Hello. Is Mike Whitworth there, please?
Anne Whitworth:	No, he's at a conference this week. Can I help?
David Bartlett:	Well, my name's David Bartlett. I met Mike last month at a conference in Belgium and he asked me to phone him when I was in London about a possible joint project.
Anne Whitworth:	Can I give him a message?
David Bartlett:	Yes, please. As I said, I'm in London this week, and my mobile number is 07700 900004. I'm leaving for Belgium again on Friday evening, so it would be good if he could call before then. Could you ask him to call me?
Anne Whitworth:	Yes, OK. Could you give me your name again?
David Bartlett:	Sure, it's David Bartlett – B-A-R-T-L-E-T-T.
Anne Whitworth:	Can you repeat the number, please?
David Bartlett:	It's 07700 900004.
Anne Whitworth:	That's fine. I've got that. I'll ask him to ring you when he gets back tomorrow evening. Bye.
David Bartlett:	Thanks. Bye.

2

Receptionist:	Western Textiles, good morning.
François Bertrand:	Hello. Is Bob Harrison there, please?
Receptionist:	I'll see if he's in the office. Who's calling?
François Bertrand:	François Bertrand.
Receptionist:	Please hold the line, I'll see if I can transfer you … . Sorry, he's in a meeting at the moment, I'm afraid. Can I help you?
François Bertrand:	Well, I met Bob Harrison in Spain last week at the Madrid trade fair. He suggested I should call him this week. When will he be free, do you know?
Receptionist:	I'm afraid I don't know. Shall I ask him to call you as soon as he can?
François Bertrand:	Yes, please, that would be good.
Receptionist:	Could I have your name again, please?
François Bertrand:	Yes, it's François – F-R-A-N-C-O-I-S Bertrand – B-E-R-T-R-A-N-D.
Receptionist:	Thank you. And your phone number?
François Bertrand:	Yes, I'm back in France now. My number is 39 46 57 93, and I think the code from the UK is 0033, then 1 for Paris.

Receptionist: Right, can I confirm the number – 0033 1 39 46 57 93.
François Bertrand: That's right.
Receptionist: I'll ask Bob Harrison to give you a ring as soon as he's free.
François Bertrand: Thank you very much. Goodbye.

Task 5

Richard Dawson: Richard Dawson speaking. Can I help you?
Hannah Booth: Oh, hello Richard, it's Hannah Booth. How are you?
Richard Dawson: Fine thanks, but busy as usual.
Hannah Booth: Sorry to bother you, I've just got a quick question. Could you give me the company name and the phone number of that person you mentioned last week? You told me about a woman who runs an import/export office in Taiwan. Do you remember?
Richard Dawson: Yes, I do. She's called Carla Parker. Hmmm … I don't know her phone number offhand, but I can look it up for you. I can't do it right now, but I could call you in about an hour. How long are you going to be in the office?
Hannah Booth: I'll be here till about six.
Richard Dawson: OK, well I'll call you before then. Talk to you later.
Hannah Booth: That's great. Thanks. Bye.

Task 6

Receptionist: Good morning. Motor Systems UK. Can I help you?
Nick Sheridan: Good morning. I'm phoning from Star Cars International. I'd like to speak to someone about an order.
Receptionist: Right. I'll put you through to the Customer Services Department.
Nick Sheridan: Thank you.
Mark Wheeler: Mark Wheeler speaking. Hello.
Nick Sheridan: Good morning, Mr Wheeler. This is Nick Sheridan from Star Cars International. I'd be grateful if you could bring our order forward, as we need the parts more urgently than we thought. Can you help?
Mark Wheeler: Possibly … can you give me the order number, then I can check?
Nick Sheridan: Er … yes, it's 83952/026.
Mark Wheeler: Hold on a moment while I get it up on my screen. Oh, yes, I see – 83952/026. You ordered 60 QP pump motors and a series of spare parts.
Nick Sheridan: That's right. Would you be able to bring forward the delivery date to October?
Mark Wheeler: Er … October, and you originally wanted the order by the end of the year. That may be difficult, as we're very busy at the moment. I'll see what we can do. Can I ring you back, Mr …?
Nick Sheridan: Sheridan. Yes, please. Call me as soon as you can.
Mark Wheeler: Could you give me your phone number? Or is it the one on the order?
Nick Sheridan: Well, that's the main office number, but my direct line is 020 7600 9421.
Mark Wheeler: OK, I've got that. Good. I'll get back to you by tomorrow at the latest.
Nick Sheridan: Thank you very much. I'd appreciate anything you can do. Goodbye.

Task 7

1

Hannah Booth: Hannah Booth speaking.
Richard Dawson: Oh, hello Hannah. It's Richard Dawson, returning your call. I've got the information you wanted about the import/export person in Taiwan.
Hannah Booth: Oh, Richard, thanks so much. Sorry I disturbed you earlier.
Richard Dawson: That's OK. I was between meetings.
Hannah Booth: Oh, right. Now what's the name of the company Carla Parker runs?
Richard Dawson: It's Atlas Import and Export. That's A-T-L-A-S.
Hannah Booth: Right, I've got that. And what's her phone number?
Richard Dawson: It's 00886 7 6588 3456.
Hannah Booth: OK ... Atlas Import and Export ... 00886 7 6588 3456. That's great. I'll call her tomorrow. Thanks very much, Richard.
Richard Dawson: No problem. I hope she can help you.
Hannah Booth: So do I. Anyway, thanks again, and sorry to have bothered you. Bye.
Richard Dawson: Bye, and hope to see you before too long.

2

Receptionist: Hello, Star Cars International.
Mark Wheeler: Hello. Is Mr Sheridan there, please? This is Mark Wheeler from Motor Systems UK.
Receptionist: Hold on ... he's not answering his phone. I'll try and find out where he is ...
Nick Sheridan: Hello, sorry to have kept you waiting.
Mark Wheeler: Hello, Mr Sheridan. It's Mark Wheeler from Motor Systems UK. You called yesterday about your order for pumps and spare parts.
Nick Sheridan: Oh yes, good. Have you been able to do anything about bringing the delivery date forward?
Mark Wheeler: Yes, I've checked with the plant, and we can send you the whole lot by the 20th of October.
Nick Sheridan: Oh, that's excellent. Really good. Thanks very much. Is that definite?
Mark Wheeler: Yes, absolutely.
Nick Sheridan: Well, it'll make a lot of difference to our efficiency this end. I'm very grateful to you for arranging it.
Mark Wheeler: That's fine. I'll email confirmation of all the new arrangements to you right now.
Nick Sheridan: Good, and once again, thank you very much for all your help.
Mark Wheeler: You're welcome. I'm glad we could do it for you. Goodbye then.
Nick Sheridan: Goodbye.

Task 11

1 Wallace. Could you spell that for me, please? (W-A-L-L-A-C-E)
2 Lefevre. Could you spell that for me, please? (L-E-F-E-V-R-E)
3 Schoppen. Could you spell that for me, please? (S-C-H-O-P-P-E-N)
4 McDonagh. Could you spell that for me, please? (M-C-D-O-N-A-G-H)
5 Takamura. Could you spell that for me, please? (T-A-K-A-M-U-R-A)
6 Cricchi. Could you spell that for me, please? (C-R-I-C-C-H-I)

Task 12

1

It's 11.30 a.m.

a Hello, is that Julia?
b Hello, is Fernando there, please?
c Good morning. Is Kirsten there, please?

Now it's 3 p.m.

d Hello, I rang earlier. Has Julia finished her meeting yet?
e Hello, could I speak to Fernando, please?
f Hello, is that the office Kirsten works in? Could I speak to her, please?

2

a Oh, hello. Listen – I've got a meeting starting in two minutes, so I can't talk long.
b Motor Systems UK. Good morning, can I help you?
c Mark Wheeler speaking. Can I help?
d I'm afraid she's in a meeting. Can I take a message?

2 Hold the line, please

Task 1 and 2

1

Receptionist:	Paperworks Printers, can I help you?
Renata Schatke:	Well, I hope so. I've just tried calling Jim Channon's direct line, but there's no reply. He asked me to call this morning.
Receptionist:	Well, he must be away from his desk. Hold the line, please, and I'll try to connect you. Who's calling, please?
Renata Schatke:	Renata Schatke.
Receptionist:	Sorry, could you repeat that, please?
Renata Schatke:	Yes, it's Renata Schatke – S-C-H-A-T-K-E.
Receptionist:	Right. Just a moment, please Ms Schatke … I'm putting you through to Mr Channon right now.
Jim Channon:	Hello, Jim Channon speaking …

2

Receptionist:	New Orbis Group, can I help you?
Yoshida Tokuko:	Yes, could I speak to Liz Hunt, please?
Receptionist:	Could you hold on for a moment, please? … She's not answering her phone, but I know she's in the office today.
Yoshida Tokuko:	Yes, she asked me to ring her today.
Receptionist:	Could you hold the line and I'll try and find her? Who's calling, please?
Yoshida Tokuko:	Yoshida Tokuko.
Receptionist:	Sorry, could you repeat your name?
Yoshida Tokuko:	Yes, it's Yoshida Tokuko – that's T-O-K-U-K-O.
Receptionist:	Thank you, Mr Tokuko. She's on another extension. Hold on a moment and I'll transfer you.
Yoshida Tokuko:	Thanks.
Liz Hunt:	Hello, Mr Tokuko. Liz Hunt speaking. I'm sorry you've had to wait. There are a few points we need to discuss …

Task 5

Tina White:	Hello, this is Tina White. Can I speak to Colin Rigby, please?
Receptionist:	I'm sorry, did you say you want to speak to Mr Rigby?
Tina White:	Yes, I did. Colin Rigby.
Receptionist:	Well, I'll just check my lists but I'm fairly sure there isn't a Mr Rigby working here.
Tina White:	I'm trying to speak to Mr Colin Rigby at Packard Enterprises. The number I have is 01632 587641.
Receptionist:	Well, that's our number, but this is Packard Electric. You've got the wrong company.
Tina White:	Oh, thank you. I'll have to check the number in the file again. Sorry to have bothered you.
Receptionist:	That's all right. Goodbye.
Tina White:	Goodbye.

Task 6

Teresa Lombardo:	Hello. Could I speak to Frank Patterson, please?
Receptionist:	Yes, could you hold on. I'll just put you through. Who's calling, please?
Teresa Lombardo:	Teresa Lombardo. I'm calling from Italy.
Receptionist:	No problem, Ms Lombardo. One moment, please.
Frank Patterson:	Frank Patterson speaking.
Teresa Lombardo:	Hello, Frank, it's Teresa Lombardo.
Frank Patterson:	Hi. How are things?
Teresa Lombardo:	Fine, thanks. What about you?
Frank Patterson:	Good, thanks. How can I help?
Teresa Lombardo:	It's about the May consignment – I wanted to warn you that it was a bit late getting to the container terminal, so it'll be a few days late arriving at your end.
Frank Patterson:	But will it definitely get here before the end of the month?
Teresa Lombardo:	Oh, yes, definitely. I've had that confirmed.
Frank Patterson:	Fine. That's no problem then.
Teresa Lombardo:	The other thing to mention is that the second container has the spare parts you ordered.
Frank Patterson:	Right, I'm glad to hear that. And I imagine the documents are on the way too?
Teresa Lombardo:	Oh yes, of course, as usual.
Frank Patterson:	Good. Is there anything else?
Teresa Lombardo:	No, that's all I wanted to confirm. Will you be coming over here soon for a visit?
Frank Patterson:	It doesn't look like it. I'm too busy, and as things seem to be going smoothly, I don't think I need to come this half of the year.
Teresa Lombardo:	Well, let us know, and we'll look forward to seeing you again before too long.
Frank Patterson:	Same here. Thanks for calling. Bye for now.
Teresa Lombardo:	Bye.

Task 7

Welcome to Central Insurance Services. In order for us to deal with your call as quickly as possible, please select one of the following options.
If you are calling to report an accident or to make a claim on your motor insurance policy, please press 1 now.
If you are calling regarding a query, a change or the renewal or cancellation of your motor insurance policy, please press 2 now.
If you are calling regarding your direct debit payments or other payments, please press 3 now.
For all other queries, please hold, and wait for a customer service adviser to assist you.
If you would like to hear the menu again, please press 4 now.

Task 11

1. When did you last see her?
   On the 9th of July, 2002.
2. What date did you start working here?
   On the 17th of September, 2001.
3. When's the meeting?
   On Wednesday, June the 12th.
4. What's your date of birth?
   It's the 7th of December, 1983.
5. What's the date of the presentation?
   Tuesday, the 25th of April.
6. When was the contract renewed?
   On February 11th, 2003.
7. When are you going on holiday?
   On Thursday, March the 15th.
8. What's the date on the document?
   The 29th of August, 1999.
9. When was the product launched?
   On the 10th of May, 2000.
10. When does the licence expire?
    On the 21st of October, 2012.

Task 12

1. Select Services, can I help you?
2. I'm afraid Liz Hunt is off sick today. Can I help?
3. Frank Patterson speaking. How can I help?
4. Hello, I wanted to speak to someone in the Human Resources department.

3 Making enquiries

Task 1 and 2

1

Menu: Welcome to Capital Investment Services.
If you would like to discuss fund management, please press 1 now.
If you wish to buy or sell shares, please press 2 now.
If your enquiry is to do with foreign exchange, please press 3 now.
If you have a general enquiry, please press 4 now.
Thank you. We will connect you with a Financial Consultant shortly …

Daniel Evans: Sorry to keep you waiting. My name's Daniel Evans. How may I help you?

Anna Woods: Hello, my name's Anna Woods. My investor reference is CIS 698 AW.

Daniel Evans: Thank you. Let me just get your details up on my screen. CIS 698 AW, did you say?

Anna Woods: Yes, that's right.

Daniel Evans: Fine, how can I help you Ms Woods?

Anna Woods: I'd like you to buy some shares for me in a company called Bioworld. I think I'd like 500 shares.

Daniel Evans: OK, I'll look into that now and call you back with the details. Could you give me your number, please?

Anna Woods: Yes, it's 01632 639404. I'll be out for about two hours this afternoon, so please call me before 3.30 if you can.

Daniel Evans: Yes, I'll certainly call before then.

Anna Woods: That's fine. Thank you.

Daniel Evans: Goodbye, Ms Woods.

2

Annabel Davies: Globe Travel Agency. Good morning. My name's Annabel Davies. How may I help you?

Dominic Lafontaine: Hello. I'd like to make a reservation for three people on a flight from London to Sydney next month.

Annabel Davies: Yes, of course. Could I have your name and phone number before we start, please?

Dominic Lafontaine: Yes, it's Dominic Lafontaine – L-A-F-O-N-T-A-I-N-E, and my office number is 01025 265265.

Annabel Davies: Thank you. Now, how many people are travelling, and what are your dates?

Dominic Lafontaine: Three of us are going, and we have to leave London on the 11th of June, returning on the 30th. We don't really mind if the flights are not direct.

Annabel Davies: I see. Well, I'll just check. Now … British Airways and Qantas flights are direct. Ah … I see that if you want to return on the 30th of June the price with BA is £895, but if you could come back on the 28th of June the return fare would be £650.

Dominic Lafontaine: And what about Qantas?

Annabel Davies: Well, mmm … it looks rather similar with them too. There seems to be less availability with a return date of the 30th of June. I know that Virgin Atlantic flights stop over in Kuala Lumpur, and they may have more availability. Shall I look into it and give you a ring back once I've got all the details?

Dominic Lafontaine:	Oh, that would be good, yes, please.
Annabel Davies:	Let me just confirm your number. I've got 01025 265265 – Dominic Lafontaine.
Dominic Lafontaine:	That's it. I look forward to hearing from you soon.
Annabel Davies:	Yes, I'll call back within an hour. Goodbye.

Task 5

Welcome to the Riverbank Cinema line. To hear film showing times or to book tickets, please stay on the line. If your enquiry is not related to film information or advanced booking, please phone the cinema direct on 020 7946 0001. Please press the star button on your telephone twice now.
Please select one of the following options.
To listen to film showing times, press 1.
To make an advanced booking, press 2.
To listen to ticket price information, press 3.
To listen to directions to the cinema, press 4.
To listen to general information, press 5.
Adult tickets: standard £6.50, superior £7.50.
Students and senior citizens, all day, every day: standard £5.50, superior £6.50.
Children under 15 years of age, all day, every day: standard £5.20, superior £6.20.
A family ticket is available for two adults and two children or one adult and three children for £17.00.
If you would like to hear the options again, please press 0.

Task 6

Receptionist:	Eastern Computers. Good morning.
Maggie Redwood:	Hello. Can I speak to Takumi Kiyama, please?
Receptionist:	Hold the line while I connect you, please …
Takumi Kiyama:	Hello. Takumi Kiyama speaking.
Maggie Redwood:	Hello, this is Maggie Redwood. How are you?
Takumi Kiyama:	Oh, hello Maggie. I'm fine, thanks, and you?
Maggie Redwood:	Just fine, thanks. Now, I'm calling because I'd like you to give me a price, please.
Takumi Kiyama:	Of course. What is it?
Maggie Redwood:	I'm trying to do some costings, and I need to know how much your XJ 33 power supplies are. I've looked at your website but I can't find the information there.
Takumi Kiyama:	It depends how many you would be ordering.
Maggie Redwood:	The first order would be for 1,000.
Takumi Kiyama:	Hold on for a moment, while I get the prices up on my screen … yes, nearly there … yes, for 1,000 the price is $29.50 each. That's not much more than the XJ 25 that you're using at the moment.
Maggie Redwood:	Yes, that's true. Anyway, I'm still doing the costings now, so I'm not quite ready to place an order yet. I'll be in touch when I'm ready to make a firm order.
Takumi Kiyama:	Good. Now while you're on the phone, can I tell you about one of the products we've been developing, which you might find very useful? The XJ 44M is a power supply for use in notebook computers.

Maggie Redwood:	Sounds interesting! Can you send me a sample for testing then?
Takumi Kiyama:	I'd be glad to. I'll confirm this by email later. Is that all for now?
Maggie Redwood:	Yes, I think so. Bye.
Takumi Kiyama:	Thanks for calling. Bye.

Task 7

Receptionist:	Good morning, Bebbington Porcelain. How may I help you?
Eva Frei:	Good morning. I'd like to speak to the sales director, please.
Receptionist:	Of course, I'll just put you through. May I ask who's speaking?
Eva Frei:	Yes, it's Eva Frei, from the International Shop in Berlin.
Receptionist:	Thank you, Ms Frei, I'm putting you through to Mr Corbett now ...
Ben Corbett:	Hello, Eva, how are you?
Eva Frei:	Fine thanks, and you?
Ben Corbett:	Generally good, but we're busy.
Eva Frei:	So are we, I'm glad to say. That's why I'm calling. I've had an enquiry from a big conference centre here, who want to replace some of their tableware. They've already done extensive research of their own, and have looked at other comparable lines from your competitors and have decided what they want. They're now comparing prices and looking for good discounts for a large order. I think that if we want their business we'll have to offer them very favourable terms.
Ben Corbett:	Yes, I see what you mean. What line are they thinking about?
Eva Frei:	They particularly like the Blue Room Collection.
Ben Corbett:	Right. Now what sort of quantity are you talking about?
Eva Frei:	They'd like the tableware service for 200 people – and of course they'd place subsequent orders with you when they want to replace more of their tableware.
Ben Corbett:	Yes, I see. Sounds good. Well, as you're one of our most important stockists in Germany, we'd be able to offer you a discount of 7.5%. Of course, I'm sure you understand that these terms are special for this particular order, and for all orders from your other customers we'd have to revert to our normal discount of 5%.
Eva Frei:	Yes, I understand that. Now can I confirm that the price of a complete boxed tableware set, with six of everything, is 120 euros?
Ben Corbett:	Yes, that's six each of the soup bowls, side plates, dinner plates, and cups and saucers. It doesn't include pasta dishes, dessert bowls or any of the serving dishes, of course.
Eva Frei:	No, I realise that. I'll have to get back to them about all the other pieces they will need. I've got all the information in your catalogue. Would you offer me the same terms for orders on those too?
Ben Corbett:	Yes, we would offer the same discount that I've just quoted.
Eva Frei:	Right, well that's very helpful. I'll get back to you once I've had a meeting with them to discuss the terms. I'm optimistic that they'll place the order within the month. Could you email me the terms of the discount you quoted, just for the record, please?
Ben Corbett:	Yes, of course, we'll do that today. And I hope your meeting with the conference centre managers goes well. Thanks for the call, Eva.
Eva Frei:	Thanks, bye.

Task 11

1 How do you spell her surname?
  T-I-P-H-A-I-G-N-E.
2 When's he arriving?
  The estimated time of arrival is 10.25 a.m.
3 How do you spell the name of the building?
  K-E-U-M-S-U-N-G.
4 When did you say the meeting will be held?
  At 2.30 p.m. on the 17th of July.
5 Could you tell me the price of a single room, please?
  A hundred and thirty euros.
6 How much do you charge for postage and packing?
  Three pounds seventy-five.
7 When do you need the report?
  As soon as possible.
8 What do we need to send you?
  The form and a stamped addressed envelope.

Task 12

Choice Travel. Can I help you?
When do you want to go?
And how long do you want to stay?
Well, there are some very early flights which don't cost too much. Do you mind what time you leave in the morning?
There's a flight at 5.20 on the 6th of April. Is that too early?
Then there's a flight back on the 13th of April at 6.30 a.m. Would that be OK?
The return fare on those flights is £180. How does that sound?
Shall I book those for you then?

Unit 4 Placing an order

Task 1 and 2

1

Taxi service:	Fast Taxi Service. Can I help you?
Maria Penella:	Yes, please. I'd like to book a taxi at about a quarter to nine, please.
Taxi service:	Sure. Where are you calling from?
Maria Penella:	I'm at Etienne's Restaurant in North Michigan Avenue.
Taxi service:	OK, and where do you want to go?
Maria Penella:	I'm going to the airport, because I'm catching a flight back to Milan tonight, but I need to stop on the way to pick up my luggage from my hotel.
Taxi service:	Yes, which hotel?
Maria Penella:	It's the Lincoln Hotel.
Taxi service:	What time's your flight?
Maria Penella:	I need to be at the airport at 10.00.
Taxi service:	That should be OK. Can you tell me your name, please?
Maria Penella:	Yes, it's Maria Penella.

Taxi service:	Right, Ms Penella, there will be a car outside the restaurant at 8.45. Please wait outside the restaurant.
Maria Penella:	Thanks very much.
Taxi service:	Thanks for your booking. Bye.

2

Mail order:	Ultra Clothing. Tim speaking, how may I help you?
Jane Chapman:	I'd like to place an order, please.
Mail order:	Certainly. Have you got a customer reference number?
Jane Chapman:	Probably, because I've ordered from you before. Where do I find it?
Mail order:	It'll be on the front of the catalogue, above your name and address.
Jane Chapman:	Found it – it's UC6845/547. Does that sound right?
Mail order:	Yes, bear with me while I find your details. Can you confirm your postcode, please?
Jane Chapman:	Yes, it's BA2 6PS.
Mail order:	And are you Ms Jane Chapman, at 61 London Road?
Jane Chapman:	That's right.
Mail order:	Now, what would you like to order?
Jane Chapman:	It's on page 44 of the catalogue, the cycling gloves, reference HG5610.
Mail order:	What size would you like?
Jane Chapman:	Large, please.
Mail order:	OK. Would you like to order anything else?
Jane Chapman:	Not for the moment, thanks.
Mail order:	That's £17.50 including post and packing. Can you give me your credit card details, please?
Jane Chapman:	Yes. My number's 4929 4750 4213 9771, expiry 02/05.
Mail order:	Thank you. That's just going through … .Your order should be with you in five working days.
Jane Chapman:	That's great. Thanks. Bye.

3

Recorded voice:	Central Office Supplies. Thank you for calling our automated ordering line. If you would like to order a copy of our latest brochure, please press 1 on your phone now. If you are already a customer and would like to make an order for stationery supplies, please press 2 now. For us to process your order, you will be asked to key in your customer number first. Please key in your customer number now, followed by the star button. Thank you. Please key in the reference number of the item you wish to order, followed by the star button.
Customer:	5736, star.
Recorded voice:	You have ordered item 5736. Now key in the quantity you require, followed by the star button.
Customer:	25, star.
Recorded voice:	You have ordered 25 of this item. If you have completed your order, please press 5. If you wish to continue ordering, please press 3. Thank you. Please key in the reference number of the item you wish to order, followed by the star button.
Customer:	4975, star.
Recorded voice:	You have ordered item 4975. Now key in the quantity you require, followed by the star button.

Customer: 30, star.
Recorded voice: You have ordered 30 of this item. If you have completed your order, please press 5. If you wish to continue ordering, please press 3.
Customer: OK ... That's all I want.
Recorded voice: Thank you. We have registered your order and you should receive it within five working days. The amount will be added to your company account at the end of the month. Please make a note of your order number now. It is COS9321/019. If you need to hear your order number again, please press 0.

Task 5

Martha Wong: Blooming Flowers, Martha speaking. Can I help you?
James Elliott: Yeah, I'd like to order some flowers to be delivered to my home address on Friday, please.
Martha Wong: Yes, that's no problem. What sort of flowers would you like?
James Elliott: Well, I suppose it depends on the prices. How much would a large bouquet of mixed spring flowers cost?
Martha Wong: Prices for a mixed bouquet start at $45.
James Elliott: OK, and how much would a big bunch of red roses cost?
Martha Wong: It depends on the number you want, of course. Roses are $5 a stem.
James Elliott: I see. Well, in that case, I'd like to order 15 red roses, please.
Martha Wong: Was that 15?
James Elliott: Yes, that's it.
Martha Wong: Right. Can I have your name, please?
James Elliott: It's James Elliott.
Martha Wong: And who would you like them to be sent to?
James Elliott: To Caterina Santiago, please.
Martha Wong: And what's the address?
James Elliott: 43 Pennsylvania Avenue, Bloomington.
Martha Wong: And when would you like them delivered?
James Elliott: They should be there before 6 p.m. on May 12th, please.
Martha Wong: Fine. And would you like to include a message on the card?
James Elliott: Yes, can you please put: Happy Birthday, all my love, J.
Martha Wong: That's fine. Now how would you like to pay?
James Elliott: Credit card. My number's 4929 8157 0984 5152, expiration is 05/05.
Martha Wong: Thank you, that's just going through ... that's fine. Thank you for your order.
James Elliott: Thanks. Bye.

Task 6

Jennifer Sato: Jennifer Sato speaking.
Serge Duval: Hello, Jennifer, it's Serge Duval, from BGX Computers.
Jennifer Sato: Hello, Serge, how nice to hear from you again. How can I help?
Serge Duval: Well, you'll remember we spoke the other day about the price you could give me for 1,000 CM 25 hard drives.
Jennifer Sato: Yes, of course.
Serge Duval: Well, we're ready to place the order now, but instead of 1,000 we now need 1,500 of them, and we need them really quite urgently.

Jennifer Sato: I see. I just need to check on the stock situation. Hold on ... yes, that's no problem, we've got plenty in stock. We can dispatch them later this week and send them by air freight, so you should have them early next week.
Serge Duval: Good. The other thing I wanted to discuss was the price. You said $89 apiece, didn't you?
Jennifer Sato: Yes, I did.
Serge Duval: But that was the price you gave me for 1,000. I was hoping you could reduce it for 1,500.
Jennifer Sato: Well, I can offer you $88 each, provided the account is settled immediately by bank transfer rather than waiting for your monthly account to be paid.
Serge Duval: OK, that sounds reasonable. I can arrange that now. Does that include insurance and delivery by air freight?
Jennifer Sato: Oh yes, it's all included.
Serge Duval: Right. In that case, it's a firm order.
Jennifer Sato: Good. I'll send you an email now with confirmation of the details of the order and the amount to be transferred. And I'll make sure the order is dispatched as fast as possible.
Serge Duval: Thank you very much.
Jennifer Sato: OK. I'll confirm everything now. Thanks for calling. Bye.

Task 7

Alessandra Tauzia: Terracotta Italiana. Alessandra Tauzia speaking. Can I help you?
Brigitte Schmidt: Oh, yes, hello Alessandra, it's Brigitte Schmidt calling from the Edelweiss Garden Centre in Zurich.
Alessandra Tauzia: Hello. How are you?
Brigitte Schmidt: Fine, thanks. I'm phoning because I was just about to send an email with a repeat order for several terracotta pots, when I realised I needed to ask a question about one of the pots in your new catalogue before I go ahead and order it.
Alessandra Tauzia: Which one?
Brigitte Schmidt: It's the 75 cm pot on page 39, item number CC75. It wasn't in last year's catalogue, was it?
Alessandra Tauzia: No, it's part of a new range we've got in from a Venetian company. What do you want to know?
Brigitte Schmidt: Well, is it guaranteed frost-proof? You know how cold it can get here – it would be terrible if they cracked in frost.
Alessandra Tauzia: Hmm..., I don't know about that, actually. I'll have to contact the company and check that out for you. I'm not sure whether they guarantee their pots that size against frost. We haven't dealt with this company before, so I'll have to come back to you on that.
Brigitte Schmidt: Well, I need to know before I order any. How long will it take to find out?
Alessandra Tauzia: I'm not quite sure, but I'll get back to you in a couple of days. Do you want to make the main order now?
Brigitte Schmidt: Yes.
Alessandra Tauzia: OK. Go ahead.

Brigitte Schmidt:	Well … we'd like 100 25 cm pots, reference number AZ25, 120 35 cm pots, reference number AZ35, 150 40 cm pots, reference number AZ40 and 175 50 cm pots, reference number AZ50.
Alessandra Tauzia:	Fine, I've got all that.
Brigitte Schmidt:	They all sold very well last season, and we've already had enquiries from customers who want more.
Alessandra Tauzia:	Great. Now this order will be dispatched on the 21st of March, and will be sent by road in the usual way. I'll be able to confirm a delivery date in a couple of days, when I get back to you about the large pots. How does that sound?
Brigitte Schmidt:	Fine. Thanks very much.
Alessandra Tauzia:	Thanks for the order, and I'll be in touch again by Thursday. Bye.

Task 11

1. What's the dialling code for Spain?
   It's 0034.
2. Have you got the item reference number?
   Yes, it's CH5067/39.
3. What's the website address of Cambridge dictionaries?
   It's dictionary.cambridge.org
4. Do you know their postcode in Vienna?
   Yes, it's A–1010 Wien.
5. Do you know what his phone number is?
   Yes, it's 0082 2 7844076.
6. What's the BBC Radio 4 website address?
   It's bbc.co.uk/radio4
7. What's the insurance policy number, please?
   It's 4381869E/06.
8. Can you give me her email address?
   Yes, it's floriane@pondnet.com
9. What's your credit card number?
   It's 5797 4132 6581 2976.
10. I need to know the invoice number.
    It's KL7954–326.

Task 12

Super Tents. Richard speaking. How may I help you?
Certainly. Could you give me the reference number from the catalogue?
OK. Was that XD–4765?
And what price have you got?
OK. Can I have your name, please?
Thanks, and your postcode, please?
That's Greenwood Road, Wimbledon, isn't it? Can you give me the number?
And what's your credit card number and expiry date, please?
That's fine. Thank you for your order. Goodbye.

5 Bookings and arrangements

Task 1 and 2

1

Travel agent:	Choice Travel. Beth speaking. How can I help you?
Mike Wilkins:	Hello. It's Mike Wilkins. I rang last week asking for information about hotels in Barcelona, as I've got to arrange a trip next month.
Travel agent:	Yes, I remember. Have you decided which hotel you'd like to stay in?
Mike Wilkins:	Well, I liked the sound of two of the hotels you told me about – the Hotel Reale and the Hotel San Lorenzo.
Travel agent:	Oh, yes, I remember.
Mike Wilkins:	Can you tell me a bit more about the hotels now, as I'd like to make the booking today?
Travel agent:	Yes, of course. I'll just find the details – I haven't actually been to either of them myself. Here we are. Well, it says the Hotel Reale is very modern and has excellent views over the city and the nearby Mediterranean.
Mike Wilkins:	That sounds quite nice. Is it central?
Travel agent:	Well, it's further from the city centre than the Hotel San Lorenzo, which we also discussed, didn't we?
Mike Wilkins:	Yes, what can you tell me about that?
Travel agent:	It says it's comfortable. It's near the cathedral, so it's certainly central.
Mike Wilkins:	How do the prices compare?
Travel agent:	Let's see … yes, just as an example, the price for a double room at the San Lorenzo is about €65 less than the Reale.
Mike Wilkins:	I think I'll go for the Hotel Reale I like the idea of a view of the Mediterranean.
Travel agent:	Right. I can certainly make the booking for you. I'll need a few more details first. Can you tell me …

2

Travel agent:	Hello, is that Helga Langendorf?
Helga Langendorf:	Yes, speaking.
Travel agent:	This is Mark from Central Travel. You asked me to find out some information for you about flights.
Helga Langendorf:	Oh, yes, thank you. Have you got it?
Travel agent:	Well, I think I've got what you wanted.
Helga Langendorf:	Hold on, I just need to find my diary and a piece of paper … yes, I'm ready now.
Travel agent:	So, you'll be in Hong Kong for a week, and then you want to go to Tokyo for a week?
Helga Langendorf:	Yes, that's right.
Travel agent:	Well, there are several flights to choose from, but the best ones for you seem to be with Japan Airlines and Singapore Airlines.
Helga Langendorf:	Is there much difference in the fares?
Travel agent:	No, there's not much to choose from on price. But the times are different of course.
Helga Langendorf:	Yes, that's what I'm more concerned about. I can't leave the conference until after lunch on the 14th of June.

Travel agent:	Well, Singapore Airlines have a flight to Tokyo at 14.30. Would you be able to make that?
Helga Langendorf:	I doubt it. When's the next one?
Travel agent:	Japan Airlines have a flight at 16.00. Would that be better?
Helga Langendorf:	Oh yes, I'm sure I could make it to the airport in time to catch that one. Could you book it for me?
Travel agent:	Yes, of course, but when do you want to go back to Hong Kong?
Helga Langendorf:	Could you book me on a flight on the 21st of June, please? I'm not so bothered what time it is that day. Just book me on whatever's available. Can you make the bookings now?
Travel agent:	Yes, I'll do it now. I've already got all your details, so I'll send you an email as confirmation and we'll send out the tickets in due course.
Helga Langendorf:	Fine. Thanks very much. Bye.

Task 5

Travel agent:	Continental Express. Good afternoon. Tony speaking. How can I help you?
Joanna Page:	Hello, it's Joanna Page. We spoke last week, and you were very helpful.
Travel agent:	Oh, yes, Ms Page, I remember. You wanted to find out about flights and hotels for your journey to Boston and Chicago. Right?
Joanna Page:	That's it. Well, I've decided what I want to do now, and I'd like you to make some reservations for me, please.
Travel agent:	Fine. Go ahead.
Joanna Page:	Well, I want to leave New York on Monday July 26 and go to Boston first. Ideally I'd like a flight around noon – late morning or early afternoon would be fine.
Travel agent:	Let me just have a look … yes, there's a flight at 12.10 from Kennedy Airport to Boston. How does that sound?
Joanna Page:	That's fine. What's the availability?
Travel agent:	Oh, there are plenty of seats left on that one. Now, where do you want to stay in Boston?
Joanna Page:	The Great Eastern Hotel in Boston. Could you book me in for two nights, July 26 and 27?
Travel agent:	Sure. What sort of room do you want?
Joanna Page:	Oh, a single room with a shower would be fine, I think.
Travel agent:	OK. A single with shower for two nights, July 26 and 27 …
Joanna Page:	Yes, that's it. Then on Wednesday, July 28 I need to go to Chicago. Is there a morning flight?
Travel agent:	Hold on a moment … yes, there's a flight leaving Boston at 10.30. How's that?
Joanna Page:	Sounds fine. Can you book that for me?
Travel agent:	Sure. Now when do you want to come back and where do you want to stay in Chicago?
Joanna Page:	Well, I've got friends in Chicago who I'll stay with, which is good, as I don't know how long I need to be there. As long as my ticket is open for my return to New York, I can make the reservation from Chicago, can't I?
Travel agent:	Yes, that's no problem. There are plenty of flights every day from Chicago back to New York. Is there anything else I can do for you now?

Joanna Page: No, that's it, as long as you email me all the details and confirmation of the reservations. That's all I need until you send me the tickets.
Travel agent: Sure. You'll be hearing soon. Thanks for making the reservations with us. Bye.

Task 6

Flavia Bellini: Good morning. Could I speak to Louis Gasquet, please?
Louis Gasquet: Speaking.
Flavia Bellini: Oh, good. Hello, Louis, it's Flavia Bellini. I'm calling so we can go through the details of your trip here next week.
Louis Gasquet: Fine. Is everything fixed?
Flavia Bellini: I think so. The hotel was booked some time ago, and we've now arranged all the meetings, so I just want to go over it all with you.
Louis Gasquet: Yes, and I've got something to ask you.
Flavia Bellini: Fine, but shall we start at the beginning?
Louis Gasquet: Sure.
Flavia Bellini: Well, you're coming next Tuesday, the 25th of October, on an Alitalia flight, aren't you?
Louis Gasquet: Yes, it's quite an early flight. It's AZ325 and it's scheduled to get in to Fiumicino Airport at 9.30.
Flavia Bellini: That's what I thought. I've arranged for a company car to collect you and bring you straight to our offices. The car will have the company name on it, and the driver will be waiting for you outside the main terminal. Just ring me or my secretary if there's a problem.
Louis Gasquet: That sounds fine. Which hotel am I in?
Flavia Bellini: We've booked you in to Hotel Locarno for two nights. It's where you stayed last time.
Louis Gasquet: Oh, yes, I remember the Locarno. It's not far from Piazza del Popolo, is it?
Flavia Bellini: That's right. Anyway, when you arrive here, just ask for me at reception, and then I'll come and find you and we can make a start going through things together. I've arranged to have lunch at 1 o'clock, followed by a meeting with the Sales Director at 3 p.m.
Louis Gasquet: That sounds good.
Flavia Bellini: There will be several key people at lunch with us, some of whom you've met before. And we've got a new American director who you'll meet then.
Louis Gasquet: Fine. I'll look forward to that.
Flavia Bellini: We'll then meet the Managing Director at 5.00, as I know you've got several things to discuss.
Louis Gasquet: Yes, that's important. Good.
Flavia Bellini: You'll stay at Hotel Locarno for two nights, but in fact on Wednesday the 26th we'll be out of Rome all day as that's the day I've arranged the visit to the main factory.
Louis Gasquet: Oh, yes, it's quite near Orvieto, isn't it?
Flavia Bellini: Yes, it's about 100 kilometres north, on the way to Florence. It takes about an hour and a half by car. We'll spend all Wednesday there, then we'll come back to Rome for dinner on Wednesday evening.
Louis Gasquet: And I know I've got a morning flight back here on the Thursday.
Flavia Bellini: Yes, and I'll make sure you'll be taken to the airport in good time for

your flight.

Louis Gasquet: Well, that all sounds fine. I wanted to ask whether there's going to be any free time, as there's an exhibition I'd really like to see.

Flavia Bellini: Well, that may be difficult, unless you change your flight on Thursday to a later one, so you can go to the exhibition on Thursday morning.

Louis Gasquet: Yes, I'll look into that. Don't let's worry about it for the moment. Is that it for now?

Flavia Bellini: Yes, I'm looking forward to seeing you again next Tuesday. Have a good journey.

Louis Gasquet: Thanks. Bye.

Task 7

Verena Fischer: Hotel Adlon Conference Centre. Verena Fischer speaking. How may I help you?

Jerry Kemp: Hello, it's Jerry Kemp here, from Spectrum Technodesign. I'm calling from the stand at the exhibition because I need to make some last-minute changes to the arrangements you're fixing for us at the hotel.

Verena Fischer: Right, let me just make a note of all the changes. Here we are. Starting with today, Monday, we've got 25 single rooms with shower or bath booked for Spectrum Technodesign staff for three nights, with 45 people booked in for dinner tonight at 8.00 in the Linden Restaurant. Is that right?

Jerry Kemp: Yes, we still want the 25 rooms. I think people will start arriving from about 6.30, probably. It's the numbers for the dinner tonight that have changed, and the number of rooms we need for the next two nights, as more people are arriving from the States tomorrow.

Verena Fischer: OK, I don't think that will be a problem, but let's start with the dinner tonight. How many people will there be now?

Jerry Kemp: Can you fit in another seven, please? There will now be 52 for dinner.

Verena Fischer: I'll let the restaurant know immediately. There's enough time to add the places, and I know there's space to get another table in the room. It would help to know if any of the extra people want a special diet. I've already got down that nine people are vegetarian.

Jerry Kemp: Yes, two of the new people are vegetarian too, so that's 11 now, but there aren't any special diets needed apart from that.

Verena Fischer: That's no problem. It just helps if I can tell the kitchen in advance.

Jerry Kemp: Sure. Now about the number of rooms we need for the next two nights.

Verena Fischer: Yes, you said you need more.

Jerry Kemp: Yes, four more staff members are flying in tomorrow. Can you fit them in?

Verena Fischer: Let's see … .Do they want singles with shower or bath, like all the others?

Jerry Kemp: Singles, ideally, but we may have to put them in doubles if that's all you've got.

Verena Fischer: I'm just checking … I'm afraid I can't put them in singles, but I've got two double rooms with twin beds that they could have – both rooms have a bath and shower. How would that be?

Jerry Kemp: That's fine, I'm sure. So we'll then have 25 single rooms and two doubles with twin beds for tomorrow night – right?

Verena Fischer: Yes, what about Wednesday night then? Is that the same?

Jerry Kemp: Yes, everyone will be staying then too, so we'll need the same rooms as Tuesday.

Verena Fischer: Now, you've booked a dinner for 15 on Tuesday night, again at 8 p.m. in the Linden Restaurant. Are the numbers the same for that?

Jerry Kemp: Well, in fact it's now going up to 17 because two of the people arriving tomorrow also have to go to that. None of them are vegetarians or have any special dietary requirements. Everyone else will make their own arrangements tomorrow.

Verena Fischer: What about Wednesday night?

Jerry Kemp: That's got to be a dinner for everyone again. And we've invited a few more visitors to join us, so we'll be 57 in total. I'll confirm that figure with you on the day, and I'll check whether there are any special diets. For the moment, there are still 11 vegetarians for that dinner. Is that OK?

Verena Fischer: That's fine. I've got all that, and I'll pass on the details to the restaurant now. Just call me if there are any more changes, please, won't you?

Jerry Kemp: Of course. Thanks. Bye.

Task 11

Rising tone ↗
She's French, isn't she?
You've booked the room, haven't you?
He doesn't like flying, does he?
Falling tone ↘
She's French, isn't she?
You've booked the room, haven't you?
He doesn't like flying, does he?

1 There aren't any seats left, are there? ↗
2 She's already paid, hasn't she? ↘
3 We'll have to change the booking, won't we? ↘
4 The dinner was good, wasn't it? ↘
5 The flight's on time, isn't it? ↗
6 You can ring them tomorrow, can't you? ↘
7 You liked that hotel, didn't you? ↘
8 They haven't called us back, have they? ↗
9 You'll make sure you're on time, won't you? ↘
10 You've got her mobile number, haven't you? ↗

Task 12

Grand Hotel. Good morning. May I help you?
Yes, I see. Can you tell me the name of the guest, please?
Oh, yes, I've got it. He's booked into a single room with bath. You want to change the reservation, do you?
Yes, that's fine. I'll change that. So, just to confirm – that's a single room with bath for Mr Gregor Bachmann for two nights from the March 14th. Is there anything else I can do?
Thank you for your call. Goodbye.

6 A change of plan

Task 1 and 2

1

Receptionist:	Manzini Partners. Good afternoon. How may I help you?
Emma Marsh:	Hello. Could you put me through to Robert Manzini, please?
Receptionist:	Yes, could you hold on, please. Who's calling, please?
Emma Marsh:	My name's Emma Marsh, from Marsh Consultancy.
Receptionist:	Yes, Ms Marsh ... I'm just connecting you.
Robert Manzini:	Hi, Emma, Robert Manzini speaking.
Emma Marsh:	Good, Robert, I'm glad you're there. I was worried you'd be away. It's about our appointment on Thursday morning. I'm not going to be able to make it now, I'm afraid.
Robert Manzini:	Oh, that's a shame.
Emma Marsh:	I'm sorry, I've had to agree to see a client from Korea on Thursday. She should have come last week in fact, but there was a last-minute change in her itinerary, and Thursday is the only day she's able to see me while she's over here.
Robert Manzini:	I see. Will she be with you all day?
Emma Marsh:	Yes, we've got to fit in meetings for her with several people, and I won't be able to get away at all. I know our meeting is urgent – I'm really sorry about this.
Robert Manzini:	No, that's OK, these things happen. But we need to find another date soon. Let me just look in my diary.
Emma Marsh:	I could manage sometime on Friday – how about you?
Robert Manzini:	No, we've got a corporate hospitality day, so we'll all be out of the office all day Friday.
Emma Marsh:	Sounds nice though.
Robert Manzini:	Yeah, it should be if the weather's good – it's a golf day for clients.
Emma Marsh:	Nice!
Robert Manzini:	Yeah, one of the benefits ...What about Monday – could you manage Monday at all?
Emma Marsh:	Yes, that looks good. I'll be tied up until about midday, but I could come to your office after that.
Robert Manzini:	Well, we could start the meeting over a quick lunch, couldn't we? I don't think there's time to go out, but I'll see if I can get one or two other people to join us for sandwiches in the office, and then we can get down to all the issues we need to go through straight after that. How does that sound?

Emma Marsh: Good, that sounds fine. Let's say I'll be at your office at 12.45 on Monday then.
Robert Manzini: Yes, OK ... 12.45 – I've put it in the diary now.
Emma Marsh: Thanks very much – I'm really sorry about having to change the date.
Robert Manzini: It's OK. At least we've managed to find another day without too much delay. Looking forward to seeing you on Monday then. Bye.
Emma Marsh: Bye.

2
Answerphone: Thank you for calling Heron International. The office is now closed for the weekend, and will reopen again at 8.30 on Monday morning. If you would like to leave a message, please do so after the tone.
David Harper: Hello, this is David Harper, with an important message for Amanda Walters. It's Sunday evening now, and I'm ringing to say I won't be able to come to the board meeting tomorrow afternoon at 2 p.m., because I've hurt my back, and I'm having to lie completely flat for at least the next two days – very frustrating, but I can't move.
I've just spoken to my colleague Helen Smith, who knows as much as I do about everything that will be discussed at the meeting, and she has kindly agreed to come to the meeting tomorrow afternoon in my place. She's going to go through all the paperwork tomorrow morning, and before she leaves for Heron International, she and I will talk through everything she will need to cover at the meeting. We prepared a list of proposals last week, which she'll bring with her.
Amanda, I'm really sorry about this last-minute change, but there's nothing I can do about it. I'll call again in the morning to check you've got this message. Bye.

Task 5

Dave Marshall: N & P Printers, good afternoon. Dave Marshall speaking. How can I help you?
Maria Safarini: Hello, it's Maria Safarini from East Bay Services.
Dave Marshall: Oh, hello Ms Safarini. How can I help?
Maria Safarini: Well, I'm ringing because we're having some problems with one of our printers, and I hope you can help.
Dave Marshall: Can you give me the model number, please?
Maria Safarini: Yes, it's a PX920 printer.
Dave Marshall: And when did you buy it?
Maria Safarini: Oh, we've only had it for about six months, so it's still under warranty, I'm sure. We bought three other identical printers at the same time, and there's nothing wrong with those.
Dave Marshall: What seems to be the problem?
Maria Safarini: Well, it works, so I don't think it's that a part has gone wrong or anything like that, but the quality and speed of the printing are nothing like as good or as fast as on the other three printers. This machine is slower and what it produces just doesn't look very good. The point is, we can compare it with the other printers, and it's obvious there must be something wrong with it – we just don't know what.

Dave Marshall: Well, we need to arrange a time for a service engineer to come and have a look at it.
Maria Safarini: Yes, that's what I was hoping for, obviously.
Dave Marshall: I'm just looking at our appointments diary. I'm afraid we're very busy at the moment. It looks as if the first day he could come would be next Wednesday, the 5th. How does that sound?
Maria Safarini: Is that really the soonest you can manage? I was hoping you could send someone tomorrow or the day after.
Dave Marshall: I'm afraid not. One of our service engineers is off sick at the moment, so next Wednesday is the earliest we can manage.
Maria Safarini: OK. What sort of time would he be able to come?
Dave Marshall: He could come to you as his first appointment of the day, at 8.00, if that's convenient.
Maria Safarini: Yes, that would be fine. I'll just make a note of that – 8 a.m. next Wednesday. Thanks very much. Bye.
Dave Marshall: Thanks for calling. Bye.

Task 6

1
Receptionist: Hotel Saint-Jean, good morning. How can I help you?
Cornelia Benz: Yes, good morning. I'd like to book a double room with shower, please, for the 25th to the 29th of April.
Receptionist: Certainly, madam. Could I have your name, please?
Cornelia Benz: Yes, it's Cornelia Benz.
Receptionist: Benz, did you say?
Cornelia Benz: Yes, that's right. Cornelia Benz – B-E-N-Z.
Receptionist: And you said the 25th to the 29th of April, didn't you?
Cornelia Benz: Yes, a double room with shower, please.
Receptionist: That's fine.
Cornelia Benz: Could you please make sure it's a quiet room, away from the main road, as I can't sleep when there's a lot of traffic outside.
Receptionist: Hold on a moment, please, I'll just change the room booking in that case … . Right, you'll be in a room with a balcony overlooking the internal courtyard, so it'll be very quiet.
Cornelia Benz: Thank you.
Receptionist: No problem, Ms Benz. We look forward to welcoming you on the 25th of April.
Cornelia Benz: Thank you. Goodbye.

2
Receptionist: Royal Western Hotel, good afternoon.
Marta Castellana: Hello. I'd like to make a booking, please.
Receptionist: Certainly, when would you like to come?
Marta Castellana: I'll be arriving on the 10th of June, and I'd like to stay for two nights.
Receptionist: What sort of room would you like?
Marta Castellana: A single room with bath, please.
Receptionist: I'm just checking that for you … . Yes, we can manage that. Now you said two nights from the 10th of June, didn't you? Could I have your name, please?
Marta Castellana: Yes, it's Marta Castellana.

Receptionist: Could you tell me how you spell your surname?
Marta Castellana: Yes, it's C-A-S-T-E-L-L-A-N-A.
Receptionist: Right, thank you, Ms Castellana.
Marta Castellana: I'd also like to book a table in the restaurant for dinner the first evening. Can you book it for me?
Receptionist: Yes, of course. What time would you like it for?
Marta Castellana: Oh, I think 8.00 would be fine. A table for six, please.
Receptionist: Right, that's a table for six at 8 p.m. on June the 10th. Is there anything else, Ms Castellana?
Marta Castellana: No, that's all for now. Goodbye.
Receptionist: Goodbye.

Task 7

1

Receptionist: Good morning. Tarquin Services. Claire speaking. How may I help you?
Barbara Zimmermann: Hello, Claire, it's Barbara Zimmermann. Could I speak to Henri Julien, please. I hope he's in the office today.
Receptionist: Yes, he's here, Barbara. I'll just put you through. Hold on a moment, please …
Henri Julien: Henri Julien speaking. Hello, Barbara, how are you?
Barbara Zimmermann: Fine, thanks, Henri. But you won't be very pleased when you hear why I'm calling. We may have to change the date of the annual dinner, I'm afraid.
Henri Julien: Oh, that's annoying. I've just had confirmation from everyone that they can make it for the 3rd of December. What's the problem?
Barbara Zimmermann: Well, it's infuriating. I've just had an email from the speaker we'd booked, saying she can't come. She's discovered she's been double-booked for the 3rd of December and she's got another speaking engagement that evening which she can't get out of. I know you particularly wanted her to speak after the dinner this year.
Henri Julien: Yes, I do. I've heard her speak before, and I know she'd have a lot of relevant things to say. Did she give you any alternative dates she could manage?
Barbara Zimmermann: Yes, she could do December the 5th or the 8th.
Henri Julien: Right – I'm just checking my diary. Could you ask her to make it the 8th then, I think that's best. We'll have to check that the Grand Hotel can put on the dinner that evening instead, and then tell everyone about the change. We'd asked everyone to come at 7.00, hadn't we, so let's stick with that time.
Barbara Zimmermann: Yes, starting at 7.00 seems fine to me.
Henri Julien: Well, let's hope the Grand Hotel can do it all on the 8th, and that they've got enough rooms then, too.
Barbara Zimmermann: I'll try to sort it out quickly and I'll call you later today, I hope. Bye for now.
Henri Julien: Bye, Barbara.

2

Receptionist:	Tarquin Services. Good afternoon. Claire speaking. How may I help you?
Barbara Zimmermann:	Hello Claire, it's Barbara Zimmermann. Is Henri Julien there, please?
Receptionist:	Yes, he is, Barbara. Hold on a moment ...
Henri Julien:	Hello, Barbara. I hope it's good news this time ...
Barbara Zimmermann:	Well, I've confirmed with the speaker that she can speak at our dinner on the 8th of December. She was very apologetic about the mix-up and having to let us down when we'd already made all the arrangements for the 3rd.
Henri Julien:	I should hope so!
Barbara Zimmermann:	The bad news is that I've spoken to the Grand Hotel, and they can't have us on the 8th of December.
Henri Julien:	Oh, no ...
Barbara Zimmermann:	Well, I've also called the Hotel Bristol, and they can arrange the dinner for us on the 8th of December, and they would also have enough rooms for everyone to stay there too, which is good, so I think we should go for that. What do you think?
Henri Julien:	Yes, that seems best. Thank you very much. In fact, I expect the Hotel Bristol will be just as good as the Grand. Thank you very much for sorting it all out.
Barbara Zimmermann:	That's OK. Shall I send an email to everyone telling them about the new place and date, or will you?
Henri Julien:	Would you mind doing it, as you've made all the arrangements?
Barbara Zimmermann:	Yes, that's fine. I'll confirm that it's now going to be at the Hotel Bristol, still starting at 7 o'clock. I'll copy you in, of course. I just hope everyone will be able to make the new date.
Henri Julien:	So do I. Anyway, many thanks for fixing it all. Talk to you soon. Bye, Barbara.
Barbara Zimmermann:	Bye.

Task 11

1 Which nights have you booked the hotel for?
The 9th to the 11th of July.
2 Have you got the reference number?
Yes, it's PB/9534–06.
3 How do you spell his surname?
H-E-Y-D-E-N-F-E-L-D-T.
4 What's her phone number in Turin?
It's 0039 011 864 4360.
5 Can you give me his email address?
Yes, it's wilson@transdeal.netvigator.com
6 What's his mobile number?
It's 07998 652714.
7 When are they due to arrive?
On the 17th of April.
8 Do you know how to spell Polozova?
Yes, it's spelt P-O-L-O-Z-O-V-A.

Task 12

Hello, it's Philippe Lamoine. I'm calling because I won't be able to make it for the 10 o'clock appointment we arranged on February the 2nd. I've got to go to the dentist at 9.30 and I won't be out in time to come to your office. Can we make it 12 o'clock instead?
Are you free in the afternoon?
Fine, I'll see you then. Bye.

7 What's the problem?

Task 1 and 2

1

Customer adviser: Penta Magazines. How may I help you?

Caroline Weaver: Hello, I'm ringing because I sent off my subscription for one of your magazines a couple of months ago, and I still haven't received the first copy.

Customer adviser: Oh, that's surprising.

Caroline Weaver: That's what I thought. And I know you received the payment, because the right amount was debited from my credit card on the 23rd of April according to my statement.

Customer adviser: Could I just have your postcode?

Caroline Weaver: Yes, it's BA2 6EW.

Customer adviser: Is that B for bravo, A for alpha?

Caroline Weaver: That's right.

Customer adviser: I'm just checking … is that Ms Caroline Weaver?

Caroline Weaver: Yes, that's me.

Customer adviser: And is your address 21 Glebe Crescent, Bath?

Caroline Weaver: Yes, that's right.

Customer adviser: Well, that's very strange. Our records show that the subscription form was received and dealt with on the 23rd of April.

Caroline Weaver: That makes sense because you took my money on the 23rd according to the statement. But I still haven't received the magazine – and I've been waiting for about seven weeks now.

Customer adviser: Well, I don't understand that, because it says here that you were sent the April issue.

Caroline Weaver: It may say that, but I never got it. Because the magazine is two-monthly rather than every month, I did wonder whether you were waiting for the next issue, but we're well into June now and nothing has come.

Customer adviser: The April one must have got lost in the post, I suppose.

Caroline Weaver: Possibly. To begin with, I thought you were just very slow at dealing with subscriptions, and that I would get my first magazine in June, but it still hasn't come.

Customer adviser: Well, it is a bit of a mystery, but we can sort it out. I'll adjust the records now and make sure it's sent out immediately.

Caroline Weaver: Thanks. But won't your records show that my subscription started in April, when in fact it should be June? I do want to get all six magazines in the year I've paid for.

Customer adviser: Yes, that'll be OK because I'll adjust it now to show that your first magazine is the June one. It's not a problem.
Caroline Weaver: Oh, that's good. Thanks very much.
Customer adviser: You should get it very soon. I hope you enjoy it.
Caroline Weaver: Thanks. Bye.
Customer adviser: Bye.

2
Pizza service: City Pizzas. How can I help you?
Harry Cox: Hello. My name's Harry Cox from the Downtown Studio. I'm calling because you delivered our pizzas 15 minutes ago but you haven't sent the number I ordered. We haven't got anything like enough. I hope it's not a mistake and the second part of the delivery is on the way and will arrive very soon – can you check, please?
Pizza service: Oh, that's surprising, I don't know what's happened – we normally send out all the pizzas for one order at the same time. Can you give me your order reference number, please?
Harry Cox: Hold on, I'll just try to find it ... yes, I've got it – it's 10964/32. It's really annoying because I specially called on Monday to make sure we'd have the pizzas for the office party today at 12.30 – maybe it would have been better if I'd called you this morning.
Pizza service: Well, we always record our orders carefully, and it shouldn't matter when you make the order. I'm just checking our records. I've got down that you wanted 13 pizzas, just –.
Harry Cox: Did you say 13? I ordered 30, not 13! No wonder we haven't got enough.
Pizza service: Well, the person who took your order must have misheard you – I'm very sorry about that.
Harry Cox: Yeah, I thought I'd said it quite clearly. But I suppose it's an easy mistake. What matters now is what you're going to do about it.
Pizza service: Well, you're missing 17 pizzas, aren't you?
Harry Cox: That's right – and everyone's due to arrive in about 5 minutes.
Pizza service: I'm sure we can get the right number to you quickly – you're not very far away, are you?
Harry Cox: No, we're not.
Pizza service: How about if we send you all 30 this time, so they'll all be hot.
Harry Cox: Well, that would be great. What about the 13 we have here?
Pizza service: We won't charge you for them. It's obviously our mistake.
Harry Cox: That's great. How soon can we expect them?
Pizza service: We'll get them to you just as quickly as we can – it shouldn't be more than 45 minutes. I'll take the order to the kitchen right now, and tell them it's urgent. We'll get the pizzas delivered to your office as quickly as possible.
Harry Cox: Thank you. So you'll send me the bill for just 30 pizzas, won't you?
Pizza service: Yes, I've got all the details. Thank you for your call, Mr Cox. We can only apologise for the mistake, and we hope you'll order through City Pizzas again.
Harry Cox: Thanks for taking care of it. Bye.

Task 5

1

Tony Martin: 'Fast Fax Central' Service Department, Tony Martin speaking. How may I help you?

Vera Steiner: Hello, I'm calling because I sent you my fax machine to be repaired about five weeks ago, and I still haven't got it back.

Tony Martin: Can you give me your name and the reference number you were given before you sent it to us, please?

Vera Steiner: Yes, my name's Vera, V-E-R-A Steiner, S-T-E-I-N-E-R, and the reference number is … hold on … RZ2984/W56. I sent it to you on the 28th of January, and it's now the 9th of March and I still haven't got it back. It seems a very long time – and I'm sure I was told I'd have it back within two weeks.

Tony Martin: I'm very sorry, Ms Steiner, I'm just trying to find out what's happened – I'm just checking through our records.

Vera Steiner: Also, this is the second time my fax machine has had to be repaired in five months. Luckily it's still under warranty but I'm beginning to wonder whether I bought the wrong fax machine in the first place.

Tony Martin: Well, I'm just checking. Yes, we received your fax machine on the 30th of January, and according to my records it was repaired within four days and was then ready for despatch on the 7th of February.

Vera Steiner: So why hasn't it arrived yet? It must have got lost. How did you send it?

Tony Martin: It was sent by courier.

Vera Steiner: Which courier?

Tony Martin: Quicklink. I'll have to contact them to find out what's happened. They're normally very reliable.

Vera Steiner: It's been very inconvenient not having it. If it's got lost I'm going to have to ask you to send me a replacement under the warranty.

Tony Martin: I'm sure there's some explanation. I'm very sorry about the delay. I'll call Quicklink and find out what's happened, and then call you back. Is that OK?

Vera Steiner: Yes, please do. You've got my number, haven't you?

Tony Martin: Yes, I'll call you as soon as I can. Bye.

Vera Steiner: Thanks. Bye.

2

Courier service: Quicklink Couriers, good afternoon. How may I help you?

Tony Martin: Hello, it's Tony from 'Fast Fax Central' Service Department. I'm ringing because we've got a problem with a delivery and I'm pretty sure it must be your fault.

Courier service: Oh, really, what's happened?

Tony Martin: You collected a fax machine from us on the 7th of February for despatch to a customer, and she still hasn't received it.

Courier service: But that's over a month ago.

Tony Martin: That's exactly why she's been on the phone to me, registering her complaint about the delay. Can you tell me what's happened to it?

Courier service: Can you give me the reference number?

Tony Martin: Yes, it's RZ2984/W56.

Courier service: OK, RZ2984/W56. Hold on – I'll just have to look it up.

Tony Martin: We'll have to send her a replacement if it's got lost.

Courier service: Here it is – I've got it on the screen now. Yes, you're right, we did collect it from you on the 7th of February. The customer is Ms Vera Steiner – is that right?
Tony Martin: Yes, that's it. So why didn't she receive it that week?
Courier service: Oh dear, I can see what's happened now. The driver went to her house on the 10th of February, but she wasn't in.
Tony Martin: Well, he should have left her a card saying he'd called, and asking her to contact you to arrange a convenient time. Did he do that?
Courier service: No, it doesn't look as if he did.
Tony Martin: Well, I've got an unhappy customer as a result.
Courier service: All I can say is that we're very sorry.
Tony Martin: When shall I tell her to expect her fax to be delivered then?
Courier service: Perhaps I should ring her myself, and then I can arrange the time directly with her. Would that help?
Tony Martin: Oh, yes, do that, please. I've got to ring her back now myself and I'll tell her to expect your call.
Courier service: Yes. I'll call her this afternoon.
Tony Martin: OK. Bye for now.
Courier service: Bye.

Task 6

1
Kathy Martinez: Good morning, Exhibition Organisers. Kathy Martinez speaking.
Ben Rushton: Hello, Kathy. It's Ben Rushton.
Kathy Martinez: Hi, Ben, how are you?
Ben Rushton: Just fine, thanks, and you?
Kathy Martinez: Good, thanks. How can I help you?
Ben Rushton: I'm ringing because we're coming to the exhibition in May and I need to discuss our requirements for the stand at the exhibition with you.
Kathy Martinez: Right. We've sent you all the paperwork already, haven't we?
Ben Rushton: Yes, I've got everything, and I've looked at the plans. In fact it looks as if the arrangements are all fairly similar to last year.
Kathy Martinez: Yes, most things will be the same again this year.
Ben Rushton: I just want to confirm a few things before I email the details to you.
Kathy Martinez: Sure, go ahead.
Ben Rushton: Right, we're going to book 40 square metres of space in the exhibition hall, and I'd like you to make sure we have a central unit, not a corner unit. Will that be OK?
Kathy Martinez: Yes, I'll make sure you get that.
Ben Rushton: Last year there was a pillar in the way – and I don't want to have a pillar again, as it meant we had a lot of problems with one of the stands.
Kathy Martinez: Well, there are several pillars in the hall, of course, so some people have to have one, but I'll try to make sure your space doesn't have one this year.
Ben Rushton: Thanks. As for what we want you to provide – we want a middle section with three walls, please. Will that be OK?
Kathy Martinez: No problem. Just put everything in your email to me, and I'll sort everything out for you. Which day will you be arriving?

Ben Rushton: Four of us will be coming on May the 7th to get everything set up and ready, and then everyone else will come on May the 9th, just in time for the opening of the exhibition. We're looking forward to coming again.
Kathy Martinez: Yes, it'll be good to see you. Bye for now, Ben.
Ben Rushton: Bye, Kathy, I'll email our requirements now.

2
Kathy Martinez: Good afternoon, Exhibition Organisers. Kathy Martinez speaking.
Ben Rushton: Hello, Kathy, it's Ben Rushton.
Kathy Martinez: Hi, Ben, I'm glad to hear you've arrived safely. Is everything OK?
Ben Rushton: Well, that's why I'm calling, Kathy. It's not OK at all. I'm in the exhibition hall right now, and the space we've been allocated doesn't match what I booked with you at all.
Kathy Martinez: Really? I'm very surprised. What's the problem?
Ben Rushton: Well, we've been given a space in the corner of the hall, and I specifically booked a central unit. And on top of that there's a pillar right in the middle of the space – so we can't set the stands up as we want. I'm afraid this isn't acceptable, and you're going to have to make some changes for us.
Kathy Martinez: I see. I'm going to have to think about this and find out what's gone wrong.
Ben Rushton: I think we'll have to ask for a refund too, as you can't expect us to pay the fee when so many things have been mismanaged.
Kathy Martinez: Hold on, Ben, we haven't reached that point yet. I'm hoping I'll be able to get everything rearranged so you can have the central unit I thought you'd been allocated. I'm afraid I asked a colleague to make the final allocation of the spaces, and it sounds as if she's made a serious mistake here. I should have checked what she'd done – it was my responsibility.
Ben Rushton: Well, I'd be very grateful if you could sort it all out now. We'd like to know what's happening as soon as possible as we've got a lot to do on the stands to get everything ready.
Kathy Martinez: Yes, I appreciate that, Ben. Hold on. I'm just looking at the plans now. I think it will work if we move you to a space in aisle 3 – it's got the letters HN by it. Can you see it from where you're standing now?
Ben Rushton: I'm just walking up aisle 3 now to have a look … oh, yes, that looks fine, Kathy. We'll move here.
Kathy Martinez: I'm sorry about this, Ben, my colleague can't have checked the booking sheets properly – I should have checked everything myself.
Ben Rushton: Well, these things happen, I know. But of course it's meant we've wasted quite a bit of time. I just hope we can get everything ready in time.
Kathy Martinez: I can't apologise enough for the mix-up. I tell you what, I'll send someone over to give you a hand moving everything. And I can offer you a 5% reduction on the exhibition fee.
Ben Rushton: Well, that would be a help. Thanks, Kathy.
Kathy Martinez: Ok Ben. Bye for now – and my apologies again.
Ben Rushton: That's OK now. Bye, Kathy.

Task 7

Receptionist: Regal Hotel, good afternoon. Can I help you?

Clare Webber: Yes, my name's Clare Webber, from City Management Services. We held a two-day conference at your hotel last week, and there are several things I was unhappy with. I'd like to speak to the conference manager, please, as I want to make a complaint.

Receptionist: Could you give me your name again, please?

Clare Webber: Yes, it's Clare Webber, from City Management Services.

Receptionist: I'm just putting you through, Ms Webber.

Clare Webber: Thank you.

Edward Bonner: Good afternoon, Ms Webber. Edward Bonner, conference manager at the Regal Hotel speaking. How may I help you?

Clare Webber: Hello, Mr Bonner. We met last week when we held our two-day conference at the hotel.

Edward Bonner: Yes, I remember, Ms Webber. I trust everything went well for you.

Clare Webber: Well, the conference itself was fine, but I'm ringing now because some of the arrangements made by the hotel weren't up to the standard we'd expected from the Regal Hotel.

Edward Bonner: I'm very sorry to hear that. Could you tell me what the problems were?

Clare Webber: Yes, certainly. The first two complaints were to do with the rooms. Twenty of us stayed in the hotel, and two people reported that their rooms weren't ready for our arrival and hadn't been cleaned properly. The first person found the bed hadn't been made up and there weren't any clean towels, so he had to wait while the room was made ready for him. That wasn't a major problem, but considering other things weren't done properly either, I felt I ought to tell you that I thought your standards had slipped – and that we won't be recommending any of our colleagues or other companies we work with to come to your hotel in the future.

Edward Bonner: I'm very sorry to hear that, of course. But can you tell me what the other problems were so that we can improve the standard of service?

Clare Webber: The second person complained that a full ashtray had been left in her room, even though she'd been led to believe there was a non-smoking policy in the bedrooms, and the cups and saucers were dirty, and had just been left by the previous occupant of the room.

Edward Bonner: I can only apologise. The housekeeper should have checked that all the rooms had been cleaned properly and were ready before a guest was allowed to go into them.

Clare Webber: The other thing, that was more major, as it affected us all, was the main dinner on the Thursday night.

Edward Bonner: What was the problem there?

Clare Webber: It was a working evening, so we'd deliberately asked for the meal to be served at 7.30 so we could continue with our discussions afterwards. We had drinks at 7 o'clock, then went to the restaurant at 7.30, expecting the meal to be served almost immediately. Everything was laid out for us, and it looked ready, but then we waited and waited – and the first course didn't arrive until 8.15, even though I and various other colleagues kept asking the waiters what the problem was – in fact

	we never got an explanation, let alone an apology. As a result, we didn't finish eating until 9.30, and we wasted a lot of time just waiting. The most annoying thing was that nobody seemed to be in charge – we should have been warned there was a problem.
Edward Bonner:	I'm very sorry to hear all this. I should have been told about it by the kitchen manager, but this is the first I've heard of the problem. I can only apologise on behalf of the hotel.
Clare Webber:	I don't feel prepared to authorise payment of your bill, which has arrived very punctually. It's a lot of money and we had expected a much better standard of service at your hotel.
Edward Bonner:	Our staff are normally very reliable, and most of our customers enjoy their stay here, and in fact many of them return regularly. I'm afraid you've experienced some very bad luck, and I shall make a full investigation into what went wrong. Thank you very much for giving me so much feedback – I shall do my best to ensure it never happens again, to you or anyone else.
Clare Webber:	No, I certainly hope not.
Edward Bonner:	As for the invoice you've been sent, please disregard it. I'll have a new one made out, with a deduction of 10% because of the poor service you received here.
Clare Webber:	Well, that would make a difference, thank you, Mr Bonner.
Edward Bonner:	Thank you very much for calling, Ms Webber, and I can only apologise on behalf of the hotel.
Clare Webber:	Goodbye.

Task 11

1 A half plus three quarters equals one and a quarter.
2 Four point two multiplied by three makes twelve point six.
3 Thirty-six divided by nine is four.
4 Seventeen thousand five hundred and six.
5 Seventy-eight point five per cent.
6 Three hundred and ninety-one minus sixty-two plus a hundred and forty-eight equals four hundred and seventy-seven.
7 Eleven sixteenths take away five eighths equals one sixteenth.
8 Seven point three plus twenty-nine point two is thirty-six point five.
9 Forty-three times five is two hundred and fifteen.
10 Two thousand six hundred and forty divided by eight is three hundred and thirty.

Task 12

Good morning. Angela Rusita in Customer Services speaking. How can I help you?
Oh, I'm sorry about that. What's the matter?
Can you give me the invoice number, please, and your order number?
And what seems to be the problem?
I see. I'm very sorry about that. There's obviously been a mix-up. I'll have it checked for you and I'll call you back as soon as I can.

8 Handling complaints

Task 1 and 2

1

Receptionist:	Data 5 Services. Good morning. Can I help you?
Claude Bernard:	Good morning. Could I speak to Amy Denver in the Accounts Department, please?
Receptionist:	Certainly. I'll just transfer you.
Amy Denver:	Data 5 Services Accounts Department. Amy Denver speaking. Can I help you?
Claude Bernard:	Yes, I hope so. It's Claude Bernard from BJZ Computers.
Amy Denver:	Hello, Claude, how are you?
Claude Bernard:	I'm fine, except we've got a problem. I'm calling to try to chase up an overdue payment.
Amy Denver:	Oh, really? We always process invoices for payment the same week we receive them.
Claude Bernard:	We've sent you three emails about this overdue payment, and we've heard nothing, so I thought I'd better call you.
Amy Denver:	I'm very sorry. I'll have to look into this. Can you give me the invoice number, please, and the date?
Claude Bernard:	Of course. It's invoice number BJZ 98452 and it's dated the 3rd of October.
Amy Denver:	And how much is it for?
Claude Bernard:	It's for €150,000.
Amy Denver:	OK, I'm just looking at our records – BJZ 98452, dated the 3rd of October, for €150,000. The invoice was processed on the 8th of October, so it should have gone through in the normal way.
Claude Bernard:	Can you check how you were paying it?
Amy Denver:	Yes, I can, hold on ... I can see an instruction to our bank to pay the full amount by banker's draft.
Claude Bernard:	Well, I wonder why we haven't received the money then – it doesn't make sense.
Amy Denver:	Oh no ... I've seen something here. Perhaps it does make sense. I can see the instruction, but there's no confirmation that the instruction was sent to the bank. I think it's been sitting here in the computer, but the person responsible didn't quite finish the job. I don't believe it!
Claude Bernard:	Well at least that would explain it.
Amy Denver:	Yes, I really am sorry about this. It's obviously a major slip-up. I'll send the instruction to the bank myself right now, marked 'urgent', so you should have payment by the end of the week.
Claude Bernard:	Thank you very much.
Amy Denver:	That's OK – I can only apologise on behalf of Data 5 Services. I'm going to have a word right now with the person who is responsible for the error and make sure it never happens again!
Claude Bernard:	Well, I'll leave you to it. Bye for now.

2

Receptionist:	Flyfast Airlines. Good morning. Can I help you?
Marina Donato:	Hello. My name's Marina Donato. I'm calling because my luggage wasn't on the same flight that I came here on yesterday – and I really need my things.

Receptionist:	Yes, I'm sure you do. I'll just transfer you to the office that deals with missing luggage. Could you hold on a moment, please?
Marina Donato:	Yes.
Rob Godwin:	Rob Godwin, Flyfast Airlines Baggage Department speaking. I understand your luggage has gone missing – is that right?
Marina Donato:	Yes, I flew to Edinburgh yesterday. I waited for ages after landing, only to find my suitcases weren't on the flight, and then I had to rush off to meet someone. When I rang your office at 9.00 last night all I got was the answerphone telling me to ring back this morning. It's been a real nuisance.
Rob Godwin:	I'm sorry about that. I'm sure we'll be able to trace it and get it to you as soon as we can. First of all, could I have your name, please?
Marina Donato:	Marina Donato.
Rob Godwin:	And what flight were you on yesterday?
Marina Donato:	It was flight FA 537 from Genoa, stopping over in London.
Rob Godwin:	I see.
Marina Donato:	We should have landed in Edinburgh at 6 o'clock, but the flight was late leaving London, and we didn't arrive here until 7.00.
Rob Godwin:	Right. Can you also give me the number of your baggage check, which you'll find inside your ticket?
Marina Donato:	Yes, there are two – FA 537 069431 and FA 537 069432. I've got a large brown suitcase and a smaller green overnight bag – and I really need them today.
Rob Godwin:	Thank you, Ms Donato, I'll be able to trace them now. Could I have your phone number so I can ring you as soon as we find them?
Marina Donato:	Yes, my mobile is 363 2005738 and you need to dial the code for Italy first.
Rob Godwin:	Right, I've got that. I'll ring you as soon as we've tracked down your baggage and we'll have it sent to you. Where are you staying?
Marina Donato:	I'm at the Horseshoe Hotel.
Rob Godwin:	That's fine. I'll ring you as soon as possible.
Marina Donato:	Thanks – it's really difficult not having anything at all.
Rob Godwin:	I'm very sorry for the inconvenience. I hope we'll be able to get your things to you today.
Marina Donato:	So do I. Thanks. Bye.

Task 5

Simon Cooper:	Bell-Watson Computers, good morning. Can I help you?
Bettina Seitz:	Yes, I hope so. I'm calling because I've had various problems with the computer I bought from your company six months ago and I'm not satisfied with the service I've been getting.
Simon Cooper:	I'm sorry to hear that. Can you give me your name, please?
Bettina Seitz:	Yes, it's Bettina Seitz.
Simon Cooper:	Did you say Seitz?
Bettina Seitz:	Yes, that's it. S-E-I-T-Z.
Simon Cooper:	Could I have your customer number, please, Ms Seitz?

Bettina Seitz: Yes, it's BS/009753.
Simon Cooper: Right, I'm just checking .... So what's the problem?
Bettina Seitz: Well, in fact, it's been one problem after the other. The thing I'm calling now to complain about is my monitor, which I returned to you for repair two and a half weeks ago, and which you have still – I haven't heard a word from you since the courier collected it. Can you tell me what's going on?
Simon Cooper: Can you tell me what was wrong with the monitor?
Bettina Seitz: Yes, it was OK for the first five months after I'd bought it, but then it started flickering intermittently, with purple stripes on the screen, so I couldn't work on it properly.
Simon Cooper: I see. That does sound kind of strange.
Bettina Seitz: Well, it was very annoying. I can't work without it, obviously, and I was told I'd have it back within a week. I've had to rent a monitor from a local company, so I'm paying out money all the time in order to be able to work.
Simon Cooper: I'm just having a look at the records here.
Bettina Seitz: The only reason for buying my computer from Bell-Watson Computers in the first place was because all your advertising said that your after-sales care was super-efficient – in fact the ad gave me the impression nothing would go wrong anyway – how wrong that was!
Simon Cooper: Hold on a moment. It says here that the courier collected the monitor from your address on May 15th.
Bettina Seitz: That's right – so I thought I'd get it back by about May 22nd. And it's now the June 4th and I still don't have it back.
Simon Cooper: I'm sorry about this. It looks as if there's been an error this end.
Bettina Seitz: Well, I knew that anyway, of course – but what do you mean?
Simon Cooper: We received the monitor at the repair centre on May 16th, but it still hasn't been fixed, I'm afraid.
Bettina Seitz: I don't believe it! Why not?
Simon Cooper: I'm afraid the repair centre is running behind schedule, so it's a case of everything being held up.
Bettina Seitz: Well, what are you going to do about it? I was told quite definitely that I'd have my repaired monitor back within a week. This simply isn't good enough.
Simon Cooper: I can only apologise, Ms Seitz, on behalf of the company.
Bettina Seitz: So when will I get it back?
Simon Cooper: It should be done very soon. I'll tell the repair centre to make it top priority and we'll do our best to get it delivered to you by the end of this week.
Bettina Seitz: Can I count on that?
Simon Cooper: Yes, Ms Seitz, I promise you you'll have it by the end of this week.
Bettina Seitz: Well, I really hope so.
Simon Cooper: Thank you for calling, and you'll have your monitor back soon. I hope you'll continue to use Bell-Watson Computers.
Bettina Seitz: Well, thank you for your help, and I sincerely hope I don't have to call you again about this.
Simon Cooper: Goodbye, Ms Seitz.

Task 6

Manager: Central Mini-cabs. Can I help you?

Anne Williams: Yes, I'm ringing to make a complaint about Central Mini-cabs. Are you the manager?

Manager: Yes, I am. How can I help?

Anne Williams: Well, my office ordered a mini-cab for yesterday, as I needed to travel across the city with three colleagues, and we had lots of heavy files to carry.

Manager: When did you book the mini-cab?

Anne Williams: My office made the booking on Monday, but I actually wanted the mini-cab for yesterday, Wednesday, and it didn't come. I was furious, and that's why I'm calling you now, as it wasted a lot of time and meant we were late for a very important meeting with our client.

Manager: You'll have to give me your name and the booking reference number you were given, so I can look into it.

Anne Williams: My name's Anne Williams, I'm at Abbey Consultants. Hold on – I've got the booking number somewhere here, I think … . Yes, here it is – 7861/P.

Manager: 7861/P – OK, I'm just looking at the records. It says you were booked for a six-seater mini-cab to collect you from Abbey Buildings at 11 o'clock.

Anne Williams: That's right, that's what I wanted. But it didn't come. I rang your number at 11.15 to ask where it was and someone told me it was on the way.

Manager: Who did you speak to?

Anne Williams: I don't know who it was, but it was a woman who said we should wait where we were. She also said she'd ring me back on my mobile, but she didn't, and I still haven't heard anything from her, so for all she knows, we're still waiting.

Manager: Oh, that doesn't sound good.

Anne Williams: No, it's not, and that's why I'm ringing you myself now, even though I'm extremely busy. Anyway, it still hadn't come at 11.45, so when an empty taxi drove past, we gave up on your mini-cab and went to our meeting by taxi. It meant we got to the client's office nearly an hour late, and we weren't feeling all that calm either by then. It may not sound much to you, but these are our most important clients and it made us look unprofessional.

Manager: I'm very sorry about this. As you know, we do regular business with Abbey Consultants, and I'm sure you'll agree we're normally very reliable.

Anne Williams: Yes, I know, and that's why yesterday was such a nuisance. I've already told my colleague who arranges our office transport all about it, and said I don't want to use Central Mini-cabs again – I can't afford to be let down again.

Manager: I can only apologise for yesterday's failure. I really don't know what happened, but I'll look into it as soon as we get off the phone, and I'll call you back.

Anne Williams: Well, don't ring me, ring my colleague who books the transport.

Manager: Yes, I know her, of course, it's Jane Brooks, isn't it?

Anne Williams: Yes, I just wanted to register my complaint direct to you.

Manager: As I said, I'm very sorry, but I'll deal with it personally and ring Jane Brooks as soon as I know what went wrong yesterday. Thank you for telling me about it.

Anne Williams: OK. Bye.

Task 7

Wendy Morgan: Superior Accommodation, Wendy Morgan speaking. Can I help you?

Martha Clayton: Hello, yes I hope you can. It's to do with our booking of an apartment which we fixed up with your so-called Superior Accommodation agency six months ago. Everything's gone wrong so far.

Wendy Morgan: I'm very sorry to hear that. Can you give me your name, please?

Martha Clayton: Yes, I'm Martha Clayton. We arrived in London this morning, expecting to find the apartment all ready for us.

Wendy Morgan: Hello, Ms Clayton, you're booked in to the apartment in 14 Palace Court, aren't you?

Martha Clayton: Yes.

Wendy Morgan: I know you called the office when you arrived at the airport, because my colleague has left me a couple of messages. She's not here now, I'm afraid. Can you tell me what's been going on and whether you're OK now?

Martha Clayton: Well, too many things have gone wrong so far, and I'm calling to tell you we're very unhappy with the arrangements and the booking. I want you to take care of this right now.

Wendy Morgan: Right. You'd better start from the beginning.

Martha Clayton: OK. When I made the booking online about six months ago, I understood that if we called the owner of the apartment when we landed at the airport, he would meet us at the apartment.

Wendy Morgan: That's right. What happened?

Martha Clayton: I called the number from the airport while we were waiting for our bags, and got no answer, so I had to leave a message saying we'd arrived. I tried again after we'd got the bags, about 20 minutes later – and again got no answer. That's when I called your colleague and told her we couldn't make contact with the owner of the apartment. She said that we should go to the apartment anyway. So that's what we did.

Wendy Morgan: And then what happened?

Martha Clayton: We waited outside the building for some time – with all our bags, tired from the journey. You can imagine we weren't very happy.

Wendy Morgan: No, I'm sure.

Martha Clayton: Anyway, your colleague called me again and said she'd tracked down the owner, and he was on his way with the key to let us in.

Wendy Morgan: I hope he had a good excuse for being so late!

Martha Clayton: Well, he did apologise for keeping us waiting for so long, and he let us into the apartment. He started showing us around and it didn't take very long to see that the apartment hadn't been cleaned since the last occupants left.

Wendy Morgan: Oh dear!

Martha Clayton: Exactly. We're not willing to stay here until it's been made ready for us. Considering we've already paid to stay in it for two weeks, this really isn't acceptable. You're going to have to find somewhere else for us to stay tonight, and we want to go there immediately. We've already waited long enough to get into the apartment, and now this.
Wendy Morgan: I see. I'm very sorry about all this. You may not believe it, but the owner of the apartment is normally very reliable and we've never had a problem of this sort before.
Martha Clayton: Well, that may be true, but it doesn't help us right now. We're not staying here until somebody has given the whole place a clean. I am so angry right now. I just can't believe it.
Wendy Morgan: Yes, I'm very sorry, and I can't explain it. I think the best thing is for you to get in a taxi now at our expense and come to the office so we can meet and discuss everything. We'll certainly find somewhere nice for you to stay tonight, again at our expense. I can only apologise.
Martha Clayton: All right, we'll do that. I've got the address of your office, so we'll come straight away.
Wendy Morgan: I'll see you in about 20 minutes.
Martha Clayton: We're on our way. Goodbye.

Task 11

1 How soon would you like delivery?
As soon as possible.
2 Would you mind spelling it for me, please?
T-R-I-F-O-N-I-D-O-U.
3 How much does it weigh?
23.5 tonnes.
4 When did the Shakespeare Festival begin?
On the fifteenth of July.
5 What's the postcode of the office in Hannover?
D–30161.
6 What number should I dial from England?
00971 3 7619455
7 When are they due to arrive?
The estimated time of arrival is 17.20.
8 What's the budget for the job?
2.5 million euros.
9 What rate is VAT?
VAT is 17.5 per cent.
10 How do you spell the name of your village?
P-A-U-L-H-A-G-U-E-T.

Task 12

1 Hello, I'm ringing from Beckford Services. You promised our order would be with us by last Friday, and it still hasn't arrived.
Yes, it's order number 78651. We told you how urgent the order was when we placed it. Can you tell me what the problem is?
Yes, please do. As soon as possible.

2 Hello, I'm calling because I think you've made a mistake with the invoice you've sent us. You've charged us far too much.
Yes, it's 340162, dated the 16th of November, and you've charged us €2,450!
We were only expecting to pay €1,450.
Yes, please call as soon as you can. Goodbye.

Answer Key

M	model answer (other correct answers are possible)

1 How can I help you?

Task 1

1 at a conference / Belgium / 07700 900004
2 in a meeting / Spain / 0033 1 39 46 57 93

Task 2

M 1 Caller: David Bartlett
Message: Please call him when you get back. He's leaving for Belgium on Friday evening. Mobile: 07700 900004.
2 Message for: Bob Harrison
Call from: François Bertrand
Message: Please call him as soon as you can. Number: 0033 1 39 46 57 93.

Task 3

1 speaking 2 moment 3 see 4 holding 5 mobile 6 in 7 here 8 hold

Task 4

1 b 2 a 3 b 4 a 5 b 6 c

Task 5

Message pad 3 is correct.
Richard Dawson / Hannah Booth / Name and phone number of Carla Parker's company in Taiwan / Richard Dawson
1 She runs an import/export office.
2 He's going to look it up.

Task 6

Mark Wheeler (Motor Systems UK) / Nick Sheridan (Star Cars International) / Bring the order forward / Mark Wheeler
1 60 QP pump motors and a series of spare parts
2 83952/026 3 October

Task 7

M 1 To: Hannah Booth
From: Richard Dawson
Information: Carla Parker's company – Atlas Import and Export, phone number 00886 7 6588 3456.
2 To: Mr Sheridan
From: Mark Wheeler
Information: The pumps and parts will be delivered by 20th October. He'll email confirmation of the arrangements.

Task 8

1 d 2 e 3 b 4 j 5 f 6 a 7 i 8 h 9 g 10 c

Task 9

M *Could*, *would* and *can* are all possible depending on the level of formality.
1 Could you tell me exactly what you're phoning about?
2 Can you give me your telephone number, please?
3 Could you spell your name, please?
4 Would you repeat your address, please?
5 Can you tell me when you'll be in the office tomorrow, please?
6 Would you confirm the delivery date of the order, please?

Task 10

1 Chinese	2 USA	3 Korea	4 French	5 Germany	6 Japanese
7 Spain	8 Dutch	9 Switzerland	10 Brazilian	11 Taiwan	
12 Swedish	13 British	14 Belgium	15 Saudi	16 Ireland	

Task 12

M 1

a No, I'm afraid it isn't. Julia's in a meeting. Can I take a message?
b No, I'm sorry, he's out at the moment. You can contact him on his mobile if it's urgent. The number is 07773 925586.
c I'm sorry, she's out of the office at the moment. If it's urgent you can call her on her mobile. The number is 07966 484912.
d No, I'm afraid she's still in the meeting. It should finish at 3.30. Can I ask her to call you then?
e No, he's working at home this afternoon. You can call him there if you want. The number is 6740035.
f Well, she does work here, but I'm afraid she's in a meeting at the moment. Can I take a message and ask her to call you back?

M 2

a OK, I'll call you back later. I've got a message for you. Bye.
b Yes, I'd like to speak to Mr Wheeler, please.
c Yes, can you tell me the price of QP pump motors, please?
d Yes, please. Could you ask her to call me as soon as she can. My name is [name], and my phone number is [number].

2 Hold the line, please

Task 1

1 Renata Schatke / Jim Channon / Receptionist
2 Yoshida Tokuko / Liz Hunt / Receptionist

Task 2

1 T 2 F 3 T 4 T

Task 3

1 dialled 2 extension 3 bothered 4 mobile 5 Directory
6 Internet 7 hold 8 code 9 confirm 10 appointment

Task 4

1 a 2 b 3 b 4 a 5 c 6 c

Task 5

Colin Rigby / Packard Enterprises / Packard Electric

Task 6

Teresa Lombardo / Frank Patterson / Explaining delay in delivery of consignment; spare parts in second container
1 It was a bit late getting to the container terminal.

2 Two.
3 He's too busy and it isn't necessary as things are going well.

Task 7

To renew insurance press number 2.
1 Number 3
2 Number 1
3 Hold the line and wait for an adviser to help you.

Task 8

1 h 2 d 3 c 4 g 5 b 6 f 7 a 8 e 9 i

Task 9

1 When will Ms Gonzalez be back?
2 Why hasn't the sales office called?
3 When does the manager normally arrive at the office?
4 Why have the documents been delayed?
5 What (number) do you/I dial for Directory Enquiries?
6 Where are you phoning from?
7 When would be a convenient time to ring you back?
8 Why has the meeting been postponed? What's the new date (for the meeting)?

Task 10

1 Meet Rosalia on Tuesday at 3.00 in the afternoon.
2 Send Mauro the information about the sales figures as soon as possible.
3 Can somebody go to headquarters on Thursday morning?
4 There's an urgent meeting on Monday morning about sales and stock figures.
5 Ring Gina as soon as possible – note that she'll be out after 2.00.
6 Pl ring Adriano re meeting Tues p.m.
7 HQ want info re no. used p.a.
8 Hiromi needs figs for presentation, e.g. budget figs cf. sales figs for Sept.
9 Prob c 200 @ conference in Feb.
10 NB MD away Tues–Thurs.

Task 11

1 On the ninth of July, two thousand and two.
2 On the seventeenth of September, two thousand and one.
3 On Wednesday, June the twelfth.
4 It's the seventh of December, nineteen eighty-three.
5 Tuesday, the twenty-fifth of April.
6 On February eleventh, two thousand and three.
7 On Thursday, March the fifteenth.
8 The twenty-ninth of August, nineteen ninety-nine.
9 On the tenth of May, two thousand.
10 On the twenty-first of October, two thousand and twelve.

Task 12

M
1 Oh, sorry, I must have dialled the wrong number.
2 Oh, yes, she asked me to call today. Can you give her a message and say I called, please? My name's …
3 Hello, Frank. It's about the appointment we've been trying to arrange. Would Monday be OK for you?
4 I'm afraid this is the wrong department. Hold on a moment, please, and I'll transfer you.

3 Making enquiries

Task 1

1 Anna Woods / Daniel Evans (Capital Investment Services) / Buying 500 shares in Bioworld
2 Dominic Lafontaine / Annabel Davies (Globe Travel Agency) / Flights from London to Sydney, out on 11 June, returning 30 June

Task 2

M
1 Mrs Anna Woods (01632 639404) – wants to buy 500 shares in Bioworld. Call back before 3.30 with details.
2 Dominic Lafontaine – wants prices and availability for flights from London to Sydney, out on 11 June, returning 30 June – 3 people. Call him back (01025 265265) with details of flights.

Task 3

1 lowest 2 give; order 3 gone up 4 increase 5 special 6 rates
7 charge 8 shares

Task 4

1 c 2 a 3 a 4 b 5 c 6 b

Task 5

To book tickets in advance – Press 2
To find directions to the cinema – Press 4
Prices:
Adult – Standard £6.50; Superior £7.50
Students / senior citizens –Standard £5.50; Superior £6.50
Children under 15 – Standard £5.20; Superior £6.20
Family ticket – £17.00

Task 6

1 1,000 2 $29.50 3 XJ 44M 4 XJ 25 5 As a sample for testing

Task 7

1 200 2 7.5% 3 5% 4 €120 5 No 6 Yes
7 She will discuss the other pieces the conference centre will need, and arrange a meeting with them to discuss the terms.
8 Within the month, she hopes.

Task 8

1 g 2 a 3 i 4 f 5 h 6 b 7 e 8 c 9 d

Task 9

1 Prisca Marchal said to me / told me she would give (us) an extra 2% discount for such a big order.
2 The manager asked/told Alicia to tell Pablo Lubertino we'd/they'd received his order.
3 The receptionist asked Xin Yuzhuo how he spelt / to spell his second name.
4 The Sales Manager asked/told me to tell her we'd offer them a bigger discount.
5 Mete Irmak said they'd paid the account by bank transfer on 17 October.
6 Daniel Tai asked/told Hanna Chang to check whether the figures in the file were correct.
7 Melissa Fu asked the receptionist to tell Abdullah Hassan that she'd called.
8 Kenny Liu asked his colleague if/whether the sale was due to end the following week.

Task 10

1 PA 2 etc. 3 max 4 SAE 5 asap 6 R&D 7 p.a. 8 Attn 9 CIF
10 NB 11 ETA 12 MD 13 re 14 e.g. 15 k 16 esp 17 GMT 18 info
M 19 Ask Tatiana re invoice asap 20 Cost = CIF $49K
21 18% p.a. interest 22 NB pay max €1,250
23 MD ETA 3.30 p.m. Wed 24 Send applicants SAE each

Task 11

1 T-I-P-H-A-I-G-N-E.
2 The estimated time of arrival is ten twenty-five a.m.
3 K-E-U-M-S-U-N-G.
4 At 2.30 p.m. on the seventeenth of July.
5 A/One hundred and thirty euros.
6 Three pounds seventy-five.
7 As soon as possible.
8 The form and a stamped addressed envelope.

Task 12

M Yes, please. I'd like to know about prices for flights to Madrid.
On the 6th of April if possible.
About a week.
No, I don't mind how early it is.
No, that would be OK.
Yes, that would be fine.
That's fine.
Yes, please.

Unit 4 Placing an order

Task 1

1 Fast Taxi Service / Maria Penella / To book a taxi to the airport (via the hotel)
2 Ultra Clothing / Jane Chapman / To order a pair of cycling gloves
3 Central Office Supplies / – / To order some stationery

Task 2

1 T 2 F 3 T 4 F 5 F 6 T

Task 3

1 place 2 note 3 item; catalogue 4 urgently 5 pay; account
6 invoice 7 stock 8 freight 9 repeat 10 catch

Task 4

1 b 2 c 3 a 4 b 5 a 6 c

Task 5

Order confirmation form 2 is correct.
1 F 2 T 3 T

Task 6

1 1,500 2 $88 3 $89 4 bank 5 1,000 6 Insurance and delivery by air freight

Task 7

Repeat order:
100 25 cm pots, ref. no AZ25
120 35 cm pots, ref. no AZ35
150 40 cm pots, ref. no AZ40
175 50 cm pots, ref. no. AZ50

1 CC75 2 Frost 3 In two days / By Thursday 4 21 March 5 By road

Task 8

1 e 2 b 3 f 4 h 5 a 6 d 7 j 8 i 9 c 10 g

Task 9

1 arrives 2 am/'m staying 3 leaves 4 are/'re arranging
5 is/'s confirming 6 starts

Task 10

1 delivery 2 information 3 cost 4 enquiry 5 charge 6 confirmation
7 call 8 suggest 9 reservation 10 book 11 cancel 12 quote
13 arrangement 14 translate 15 guarantee 16 flight

Task 11

M
1 It's double oh 34 / zero zero 34.
2 Yes, it's CH5oh/zero67 (forward) slash 39.
3 It's dictionary dot cambridge dot org.
4 Yes, it's A dash 1 oh/zero 1 oh/zero Wien.
5 Yes, it's double oh / zero zero 82 2 78 double 4 oh/zero 76.
6 It's bbc dot co dot uk (forward) slash radio 4.
7 It's 4381869E (forward) slash oh/zero 6.
8 Yes, it's floriane at pondnet dot com.
9 It's 5797 4132 6581 2976.
10 It's KL7954 dash 326.

Task 12

M
I'd like to order a tent, please.
Yes, it's FD–4765.
No, FD–4765.
It's £45.
Alex Harvey.
It's SW19 1QU.
Yes, it's number 74.
4954 6712 3695 3781, expiry 04/05.

5 Bookings and arrangements

Task 1

1 Choice Travel / Barcelona / Hotel Reale; Hotel San Lorenzo
2 Central Travel / Tokyo / Japan Airlines; Singapore Airlines

Task 2

1 F 2 F 3 T 4 T 5 F 6 T

Task 3

1 double 2 departure 3 airlines; convenient 4 scheduled 5 check-in
6 apartment 7 kept 8 facilities 9 arrangements; hearing 10 book

Task 4

1 b 2 a 3 a 4 c 5 a 6 b

Task 5

Message pad 3 is correct.
Caller: Joanna Page / Hotel location: Boston / Hotel name: Great Eastern Hotel / Booking dates: 26 + 27 July / Type of room: single with shower

Task 6

1 25 October 2 9.30 a.m. 3 Company car 4 Hotel Locarno 5 2
6 1 p.m. 7 Sales Director 8 5 p.m. 9 (main) factory 10 dinner
11 No, he's been before. 12 He wants to know whether there will be any free time (he wants to go to an exhibition).

Task 7

1 52 2 11 3 25 4 2 5 17 6 no/0 7 25 8 2 9 57 10 11

Task 8

1 f 2 b 3 g 4 i 5 a 6 j 7 d 8 c 9 h 10 e

Task 9

1 We might visit Amsterdam on the way home.
2 The consignment should reach you at the end of the week.
3 You should get a good discount from the car company.
4 The discount will be bigger if you book more than 50 seats.
5 The reference number should be at the top of the page.
6 She will/'ll ring you before 12.00 tomorrow.

Task 10

1 They wanted to know if/whether all the arrangements had been made.
2 He asked me what the reference number was.
3 She enquired if/whether the hotel was central.
4 He wanted to know how much a double room cost per night.
5 She asked how long the conference would last.
6 He wondered if/whether he could pay by credit card.
7 She wanted to know if/whether I/we had booked the hotel.
8 He asked why I/we had changed the flight.
9 She wondered what they had done with the files.
10 He enquired what time the dinner would start.

Task 12

M Hello. We have a reservation for two nights from the 7th of March, and I'm afraid we need to change it.
Yes, it's Gregor Bachmann.
Yes, please. Can we change it to two nights from the 14th of March?
No, that's fine now, thank you.
Goodbye.

6 A change of plan

Task 1

1 Emma Marsh / Robert Manzini / Thursday (morning) / Korean visitor with change of itinerary / Meeting on Monday, start at 12.45
2 David Harper / Amanda Walters / Tomorrow (Monday) at 2 p.m. / He's hurt his back / Helen Smith (colleague) will come instead

Task 2

1 T 2 T 3 F 4 F 5 T 6 T

Task 3

1 hear; sounds 2 last-minute 3 diary; manage 4 make; tied up 5 postpone 6 convenient 7 free 8 appointments 9 meeting; paperwork 10 confirmation

Task 4

1 a 2 b 3 c 4 a 5 c 6 b

Task 5

Maria Safarini, East Bay Services / Dave Marshall, N & P Printers / Problems with printer (PX920) / (Next)Wednesday 5th, 8 a.m.

Task 6

1 Hotel Saint-Jean: Cornelia Benz / April 25–29 / Double room with shower / A quiet room
2 Royal Western Hotel: Marta Castellana / June 10–11 / Single room with bath / A table for six at 8 p.m. on 10 June

Task 7

1 3 December 2 Speaker can't come 3 5 or 8 December 4 Grand 5 7 p.m. 6 3 December 7 Grand 8 Bristol 9 8 December 10 7 11 She's been double-booked. 12 Because she's made all the arrangements.

Task 8

1 e 2 a 3 j 4 g 5 f 6 h 7 b 8 d 9 i 10 c

Task 9

1 If the hotel was/were fully booked, I would/'d find another hotel.
2 If the speaker couldn't come to the conference, we would/'d look for a replacement.
3 We would/'d postpone the meeting if our client couldn't come (to it).
4 It would be a disaster if the computers crashed.
5 If the sales figures were well below the target, Markus would resign.
6 I would/'d take the train if I missed my flight.
7 If Anne Marie forgot to go to her dental appointment, she would/'d call to apologise.
8 If I were ill that week, Catherine would stand in for me.

Task 10

1 back 2 with 3 through 4 in 5 with 6 in 7 out 8 for

1 g 2 f 3 b 4 c 5 h 6 d 7 e 8 a

Task 11

M 1 The ninth to the eleventh of July.
2 Yes, it's PB (forward) slash 9534 dash 06.
3 H-E-Y-D-E-N-F-E-L-D-T.
4 It's double oh /zero zero 39 oh/zero double 1 864 436 oh/zero.
5 Yes, it's Wilson at transdeal dot netvigator dot com.
6 It's oh/zero 7 double 9 8 652714.
7 On the seventeenth of April.
8 Yes, it's spelt P-O-L-O-Z-O-V-A.

Task 12

M Sorry, I'm afraid I'm tied up at 12 o'clock.
I'll just check. Yes, I'm free at 3.00.
Bye.

7 What's the problem?

Task 1

M 1 Caller: Ms Caroline Weaver
Address: 21 Glebe Crescent, Bath BA2 6EW
Notes: Subscription received 23 April
April issue sent but not received (lost in post?)
Need to adjust records so subscription starts in June

2 Caller: Harry Cox
Order no: 10964/32 (Monday 17 July)
13 pizzas
Delivery: 21 July, 12.30, Downtown Studio
Notes: Harry Cox ordered 30 pizzas, not 13. 13 were delivered 15 minutes ago.
Mistake made by City Pizzas.
Must now send 30 (urgently). City Pizzas won't charge for the 13 sent earlier.

Task 2

1 T 2 F 3 T 4 T 5 F 6 T

Task 3

1 serious 2 damaged; delivered 3 service 4 disappointed 5 apologise
6 mix-up 7 delay; processing 8 inconvenience

Task 4

1 b 2 c 3 a 4 c 5 c 6 b

Task 5

Vera Steiner / Fax machine sent for repair – still not received (over a month since it was ready for dispatch) / Ring Quicklink (courier) to find out reason for delay; they collected it on 7 February

1 7 February 2 10 February 3 No 4 delivery

Task 6

email 2 is correct.

1 From the exhibition hall.
2 It's in a corner and there's a pillar in the middle of the space.
3 It's Kathy's colleague's fault, but Kathy should have checked everything herself.
4 She arranges for Ben to move to a different space in the hall.
5 She offers to send someone to help him move everything; she offers a 5% reduction on the exhibition fee.

Task 7

Clare Webber (City Management Services) / Regal Hotel / (1) Room not ready: bed not made, no clean towels / (2) Room not ready: full ashtray, cups and saucers dirty / (3) Dinner: meal expected at 7.30, had to wait until 8.15 – given no explanation or apology

1 Two days 2 20
3 The late dinner (because it affected everyone and delayed their work)
4 No 5 by 10%

Task 8

1 f 2 b 3 g 4 j 5 a 6 d 7 i 8 c 9 h 10 e

Task 9

1 I'm sorry. We should have checked the rooms were ready.
2 I'm sorry. We should have warned you about the delay.
3 I'm sorry. We should have checked your order.
4 I'm sorry. We should have packed them properly.
5 I'm sorry. We should have put one in the box.
6 I'm sorry. We should have checked the envelope before sending it out.

Task 10

1 I'll have it checked for you.
2 I'll have it fixed for you.
3 I'll have them forwarded to you.
4 I'll have it sent to you.
5 I'll have them dispatched to you at once.
6 I'll have it brought down for you.

Task 11

M

1 A/one half plus/add three quarters equals / makes/ is / is equal to one and a quarter.
2 Four point two multiplied by/times three makes/equals/is/is equal to twelve point six.
3 Thirty-six divided by nine is / makes / equals / is equal to four.
4 Seventeen thousand five hundred and six.
5 Seventy-eight point five per cent.
6 Three hundred and ninety-one minus / less / take away sixty-two plus/and/add a hundred and forty-eight equals / makes / is / is equal to four hundred and seventy-seven.
7 Eleven sixteenths take away / minus / less five eighths equals/is/makes/is equal to one sixteenth.
8 Seven point three plus/add/and twenty-nine point two is / makes / equals / is equal to thirty-six point five.
9 Forty-three times / multiplied by five is / equals / makes / is equal to two hundred and fifteen.
10 Two thousand six hundred and forty divided by eight is / equals / makes / is equal to three hundred and thirty.

Task 12

M

Hello. I'm ringing from Zanda Electrics. I'm afraid I have to make a complaint.
The invoice you've just sent us isn't correct.
The invoice number is 54391, and our order number is 8451.
You've invoiced us for 50 clips, but we only ordered 15.
OK. Thanks.

8 Handling complaints

Task 1

1 BJZ Computers (Claude Bernard) / Data 5 Services (Amy Denver) / To chase up overdue payment / Instruct bank to make payment (urgently)
2 Marina Donato / Flyfast Airlines (Rob Godwin) / To find missing luggage / Flyfast Airlines to trace luggage

Task 2

1 T 2 T 3 F 4 T 5 F 6 T

Task 3

1 overdue 2 details; sort 3 nuisance 4 mistake 5 baggage; trace
6 deals; sick 7 sorry; apologise 8 confirmation

Task 4

1 c 2 c 3 b 4 a 5 c 6 b

Task 5

1 Bettina Seitz 2 BS/009753 3 monitor 4 15 May 5 repair centre 6 this week

Task 6

Anne Williams (Abbey Consultants) / Central Mini-cabs / To complain about mini-cab not arriving / Central Mini-cabs manager will look into it and call back

1 Monday 2 Yesterday (Wednesday) 3 To a meeting at a client's office
4 By taxi 5 No

Task 7

Martha Clayton / Superior Accommodation / To make a complaint about arrangements for the apartment they'd booked / Owner didn't answer the phone; they had to wait outside; the apartment hadn't been cleaned since the last occupants left

1 T 2 T 3 T 4 F 5 F

Task 8

1 d 2 e 3 g 4 i 5 h 6 c 7 f 8 b 9 j 10 a

Task 9

1 may/might/could 2 may/might/could 3 should/ought to
4 can't 5 must 6 must 7 should/ought to 8 may/might/could

Task 10

1 announcement 2 schedule 3 cooperate 4 apology
5 preference 6 state 7 complaint 8 reference 9 describe
10 prepare 11 delay 12 arrive 13 recommendation 14 please
15 transmit 16 departure

Task 11

1 As soon as possible
2 T-R-I-F-O-N-I-D-O-U
3 23 point 5 tonnes
4 On the fifteenth of July
5 D dash 3 oh 161
6 Double oh 971 3 76194 double 5
7 The estimated time of arrival is seventeen twenty
8 2 point 5 million euros / 2 and a half million euros
9 VAT [V-A-T] is 17 point 5 / 17 and a half per cent
10 P-A-U-L-H-A-G-U-E-T

Task 12

M 1 Could you give me your order number, please?
Well, I'll have to check. Can I call you back?
Yes, I will. I'm very sorry about this. Goodbye.

2 Oh, I'm sorry about that. Could you give me the invoice number, please?
And how much should it be for?
I'm very sorry about that. I'll have to see what's gone wrong and call you back. Is that OK?
I'll call you as soon as possible. I'm very sorry about this. Goodbye.

Acknowledgements

The authors would like to thank the Centre for Professional & Business English, Hong Kong Polytechnic University, and the Centre for English Language Teacher Education, University of Warwick, for assistance from colleagues and friends; Gary Hayes for contributing the US calls; John Block for help at the initial stage; Professor Veronica Smith for professional advice; and Will Capel, Sarah Almy and Tony Garside for consistent assistance and support.

The authors and publishers are grateful to the following copyright owners for permission to reproduce copyright material:British Telecommunications plc for the extract from the telephone directory on page 10, the information on page 19, the text and photograph on page 38, the payphone instructions on page 59, and the text and photograph on page 77; Good Housekeeping/The National Magazine Company for the adapted articles on pages 19 and 68; Andrew Mann Ltd for the Alex cartoons on pages 37 and 76; Washington Flyer Magazine for the article on page 28.

In the cases where it has not been possible to identify the source of material used, the publishers would welcome information from copyright owners.

The Publisher would like to thank the following for permission to reproduce photographs.

Getty Images: pages1, 10, 19, 27, 28, 36, 46, 55, 64, 67; SHOUT /Alamy: pages 11, 14, 34, 56; Corbis UK Ltd: pages 16, 36, 37.

Design by Kamae Design, Oxford

Printed in the United States
By Bookmasters